Darwin's Garden

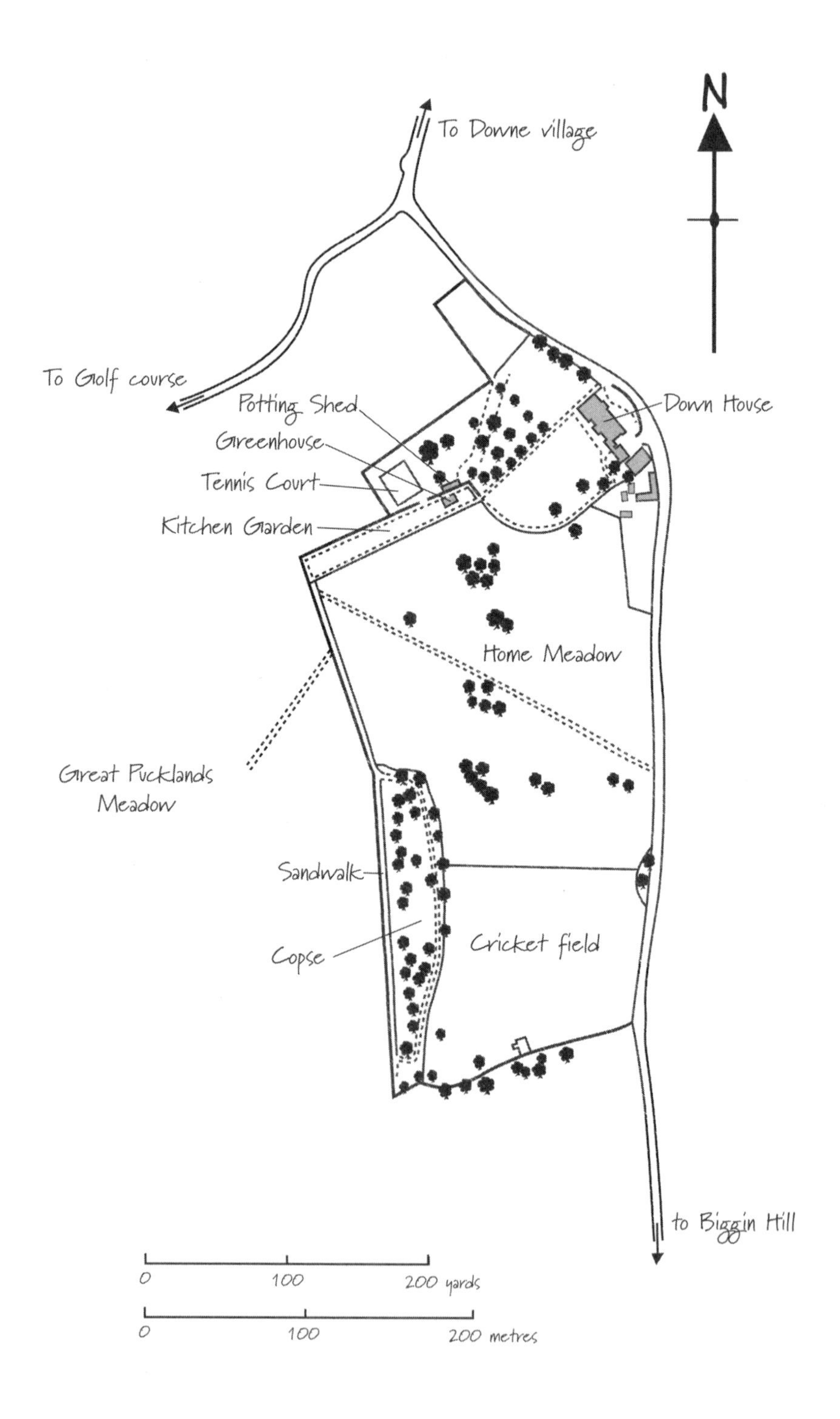
N
To Downe village
To Golf course
Potting Shed
Greenhouse
Tennis Court
Kitchen Garden
Down House
Home Meadow
Great Pucklands Meadow
Sandwalk
Copse
Cricket field
to Biggin Hill
0
100
200 yards
0
100
200 metres

Darwin's Garden

Down House
and
The Origin of Species

Michael Boulter

COUNTERPOINT
BERKELEY

First published in the UK by Constable,
an imprint of Constable & Robinson Ltd, 2008.

Library of Congress Cataloging-in-Publication Data

Boulter, Michael Charles.
Darwin's garden : Down House and the
Origin of species / Michael Boulter.
p. cm.

ISBN 13: 978-1-58243-471-1

Includes index.
1. Darwin, Charles, 1809-1882—Homes and haunts—England.
2. Darwin, Charles, 1809-1882. On the origin of species. I. Title.

QH31.D2.B7416 2009
576.8'2092—dc22
[B]

2008047237

COUNTERPOINT
2117 Fourth Street
Suite D
Berkeley, CA 94710

www.counterpointpress.com

Distributed by Publishers Group West
Printed in the United States of America

10 9 8 7 6 5 4 3 2 1

To Biddy, Tom and Alex, with love

Contents

List of Illustrations

Darwin experimenting in the potting shed at Down House. Portrait by John Collier (1850–1934). *Down House, Kent / The Bridgeman Art Library* (5512).

Alfred Wallace (1823–1913). Photograph, *c.* 1860. *Topham Picturepoint / Topfoto* (0222345)

Thomas Henry Huxley (1825–95). Lithograph, English School, 19th century. *Private Collection, Ken Welsh / The Bridgeman Art Library* (238053).

Sir Francis Galton (1822–1911). Photograph, English School, 19th century. *Archives Larousse, Paris, France, Giraudon / The Bridgeman Art Library* (179489).

William Bateson (1861–1926). Photograph, 1914. *Ann Ronan Picture Library / HIP / TopFoto* (0004224).

Sir Richard Owen (1804–1892). Engraving, English School, 19th century. *Private Collection / The Bridgeman Art Library* (135980).

Charles Darwin. Photograph, Elliot & Fry, *c.* 1875–80. *Académie des Sciences, Paris, France, Archives Charmet / The Bridgeman Art Library* (207670).

Introduction

The route from London to Down House on the south-east edge of the metropolis is a journey through recent history. From Tower Bridge there are what's left of the Victorian terraced houses along the Old Kent Road that evoke memories of slums, remnants of small industries and the bigger docks. On through Lewisham, the Broadway leads to the now tired semi-detached suburbs of the 1930s, the once proud churches and corner shops. Here is the poet John Betjeman's city suburb with its boating lake and lido, fading with the image of Britain's interwar confidence.

The road then goes up the long steady hill towards the escarpment where there were watchtowers for the unkindly doodlebug bombs that came in the last months of the Second World War. A turning to the right goes up to Biggin Hill aerodrome, now an airport, beyond the big new hospital. Just before the runway there's a crossroads with a narrow little lane that takes the traveller into the heart of Downe village. The village name is now spelt with an 'e' while the house that belonged to Charles Darwin is still plain 'Down'. Along the lane you have to be careful of the little red buses on route 146, where the bus-stops double as passing places.

Down is just over 200 feet (65 metres) above sea level near the top of a chalk hill that catches winter weather from Russia. There's still beech woodland on this acidic soil, some of the last to remain within London's M25 orbital motorway. Housing estates have taken over most of the space, the few remaining farms squashed between these signs of continuing population growth. The village still has a post office, a pub called the George and Dragon, a grocery shop and St Mary's church with its square flint tower and round spire. About two hundred years ago there were no more than fifty houses, mostly occupied by tenants working on the land. Now there's an Indian restaurant as well.

The village is surrounded on the north side by the grand estate of High Elms, until recently the family home of the local squires, the Lubbocks, who owned much of the surrounding land. The grounds and gardens of the estate are so well integrated within the landscape that they give a formal order to nature. In the mid-nineteenth century Sir John Lubbock was both a banker and a backbench Member of Parliament and so on occasion the Prime Minister, William Gladstone, called in at Downe for tea. From 1842 until his death in 1882 Charles Darwin lived on the southern edge of the village at Down House. It wasn't often that he met famous men like the Prime Minister: 'What an honour that such a great man should visit me,' Darwin wrote, his modest reaction to meeting Gladstone contrasting with his neighbour's hauteur.

As well as owning land and wielding power Sir John Lubbock was a respected astronomer and a Fellow of the Royal Society, a man who sought to explain nature objectively with laws and equations. His view of a highly ordered world was in contrast to the more flexible and uncertain attitudes of his neighbour. For these new men of the Society at the beginning of the nineteenth century, humanity had power over nature. In contrast, Charles had much greyer views with no fixed answers about the nature of science and society. For him, the secrets and complexity of life were not going to be discovered and understood so easily.

Mid-nineteenth century England was a society of strict hierarchy. The squalor of Dickens's London blotted out most of the capital's natural landscape, while at Downe beech forest covered most of the chalk upland of the North Downs. Yet stark divisions between rich and poor could be found in both town and country. The congested slums of the city had their own rural reflections – the so-called 'Good Poor': servants and labourers who worked for, or had been pensioned off by, gentry like the Lubbocks and the Darwins. The rose-covered cottages that housed growing numbers of these country folk did little to disguise the hardness of country life. It was a rigid routine that came through centuries of tradition, governed from the pulpit and the manor.

The Industrial Revolution had recently shattered this stasis, population was suddenly increasing, towns were expanding and control over the surplus human population was becoming the overbearing issue of the day. For the poor of Downe there was the threat of new workhouses or enforced migration to the colonies. As well as the possibility of civil unrest during the 1830s, there was real fear in England of another war with France. Darwin himself was to become afraid that the French army would march across his land on their way from Dover to London. He was relieved that Martello Towers had been built along the south-east coast a few decades earlier to warn off Napoleon's navy.

The political and social ferment of the western world during the 1830s also reflected the huge debate within the scientific community where the chambers of numerous societies argued about the control and meaning of life, the growth of human population and the influence of science and technology on ordinary lives. The advances in science were intertwined with the growth of overseas exploration, made possible by the new technologies of marine engineering and mapping. This enabled not only the growth of empires and the establishment of the New World, but also offered the opportunity for the exploration of new lands and the nature which inhabited it.

In the early 1840s, however, fear of what the economic future might bring still haunted English society. The cost of the Napoleonic Wars, and the rural unrest of the reform movement of the 1830s, placed most villages in jeopardy. At Downe, the local vicar, who had been living in a large square house to the south of the village, was forced to move out when his living became too hard. Down House was put up for sale. It stood empty for two years before the 33-year-old Charles Darwin came to look around. The house was big with four reception rooms downstairs and more than eight bedrooms upstairs, and large fields outside as well as two tied cottages. At £2,200 it was cheap and was available immediately. Emma, Darwin's wife, was to have their third child any week and they wanted to move out of London before then.

Charles and Emma had married in 1839 and as first cousins had known one another since childhood. Emma's father, Josiah Wedgwood, owned the famous Staffordshire pottery and involved his family in Whig politics and political reform. Emma held strong views and she was more outspoken than most women of her generation. In this she was lucky to have been influenced by many eminent figures of the early nineteenth century, such as Wordsworth, Coleridge, Byron and Sydney Smith, who were friends of the Wedgwoods and stayed with them at Maer Hall in the Midlands. The house was only 20 miles (32 kilometres) from the Darwins' home in Shrewsbury and the couple had fallen in love when Charles returned from his four-year world voyage in 1836.

It was the garden that attracted Charles to Down House. Although he thought the house was ugly it did have the potential to accommodate his growing family and servants. The garden, however, was just the place where he could play with his ideas and test his latest thoughts. As well as the landscaped plots there were orchards, three fields and plenty of space for new planting.

Away from the frightening institutions of London but close enough for the odd visit, the prospect of a new routine in this idyllic setting was too much for the young man to resist. He could be with his growing family and spend time writing to his scientific friends about the results of his experiments and the observations from his microscope work. Here, at Downe, the interactions of all that is in nature were on display, intricate processes on many scales of time and space, all around him. To sit and observe it all happening, to think about his carefully planned experiments, to be seduced by the slowly emerging secrets around him would be ideal; the garden would become his laboratory and allow him to peer ever closer into the biological minutiae of life.

Through this mix of solitary contemplation and experimentation, Darwin hoped to find the answers he was looking for.

The jumble of ideas he had brought back from South America on board HMS *Beagle* in October 1836 were beginning to make sense. It was as though all the perceptions that he had come to, drawn from his observations on the other side of the world, could be seen here following a similar pattern, linking places, environments and species. He had begun to notice these traces on his voyage and had been surprised to find others making similar observations in the conversations he had had since returning to London. But there were plenty more questions to be answered before a defensible explanation of species distribution could be put before the public. And he had many more reasons, mostly from observing the disasters of others, to be particularly cautious. Until he could gather more evidence, he locked away the notes of his tentative thoughts in a cupboard. This was one of the first things he did when he and his family moved into the house. Yet he was more determined than ever that his scientific work in the new Kentish garden would bear fruit. At Down House he would dedicate his life to finding out how species progressed. (At the time, 'transmutation of species' was how most people referred to the mysterious ways in which species changed through geological time. Subsequently, other terms have been used for evolution involving processes such as natural selection, adaptation and, more recently, genetic drift. In this book, a variety of terms will be used to describe Darwin's ideas, including the word 'evolution' itself.)

Darwin believed that scientific enquiry could unlock the secrets of evolutionary biology. He, supreme of his generation, was determined to stick to the path leading to some reasoned scientifically testable explanation of life on earth. He was, however, not the first to ask such questions, as we shall see. One of Charles Darwin's strengths was his pluralism and he was first to say that he was not an expert in any one field. At times his lack of a particular discipline worried him, making him think he had no credibility in a scientific world that was becoming increasingly specialized. This insecurity made him work the harder to win respect from the establishment, but in fact he need not have worried. After his *Journal* of the voyage of the *Beagle* was published in 1839, one reviewer wrote of the 'charm arising from the freshness of heart which is thrown over these virgin pages of a strong intellectual man and an acute and deep observer'. Darwin's financial and institutional independence, his natural choice of close friendships, and his eagerness to correspond with as many minds as he could, all made him a widely respected figure.

Today, they still play cricket at the far end of Home Meadow, the 10-acre (4-hectare) field along the edge of the lawns. Nearer the house there are flowerbeds, the Kitchen Garden and the sandy pebble path leading to a copse at the far end. The plots are still there, lovingly planted out with the same varieties the gardeners believe were used in the middle of the nineteenth century. The classic varieties of orchids, primroses and sundews are still cultivated in the long greenhouse by the orchard wall. Nowadays the vegetables from the Kitchen Garden are given to the local English Heritage volunteers who keep the place going as a museum. Charles and his wife Emma would have approved of this, for though the garden is well looked after it is still informal. The atmosphere of the Darwin family lifestyle still lives on in every nook and cranny.

Darwin's ideas have also progressed. DNA sequences, newly recognized genes that switch things on and off in developing embryos, new fossil finds and different mutants all give a grander collective impression of the whole picture of biodiversity than ever before. There are exciting trends that see everything in nature as self-organizing and interacting, with dependent species reacting to different environments, themselves becoming channels of change. They are discoveries showing up in the very organisms that Darwin brought to his garden: the orchids and primroses, earthworms and pigeons. Nonetheless it is surprising to think how the garden at Down House still stands at the centre of the science of evolution. What Darwin first sought in his Kent garden is as relevant today as it was in the 1840s.

This book tells the story of what went on in the lives and minds of the people at Down House 150 years ago. It shows the many different links between their feelings and their politics, the technology and life styles of their day. The first part of the book describes Darwin's life at Down House and the experiments he conducted in the 'laboratory' garden – from orchids, pigeons and primulas to having his children play music to worms to test their musical responses. It was here that Darwin was inspired to work out some of the subtleties of natural selection and his ideas of evolution, and here that he brought those theories to fruition in publication of his most famous book, *The Origin of Species*. The second part tells how these webs evolved through the twentieth century. Some of the reasoning led to the extinction of certain ideas, while others were adapted into the new languages of mathematical and molecular biology. It is a tale of opposing attitudes, clear facts and cloudy observations, all growing out of Darwin's attention to the detail of what was going on in the fields and garden around Down House. A recent surprise is that we can find the origins of even the newest ideas in evolution in Darwin's garden: in it lay the key to the great story of evolution not only in Darwin's time, but in our own.

Part
One

Down House

Baggage on the *Beagle*

Robert Darwin didn't write to his son very often but the few letters that did get through to the *Beagle* show an unusually warm affection: 'I got a Banana tree. I sit under it and think of you in similar shade.' The Shropshire doctor appreciated his son's interest in natural history and supported him both financially and in spirit. It must have been strange to come across banana trees in Shrewsbury during the 1830s, even though this one was in the glass conservatory belonging to a well-off middle-class doctor. Charles's father had gone out of his way to find the plant and the banana tree symbolized both Robert's love for his son and his fears for him.

Charles knew some of the detail of what was going on at home, but while he remained in touch with his father, brother Ras and sisters, and shared their values, there was a part of him that was relieved to be away from the anxieties of his family life in England. For the first time he had the freedom to set out on his own unique course of adventure. On the voyage, his life became a tale of action, explorations in new territory, interspersed with hours bent over a microscope in his cabin looking into the cells of the organisms he had found for the first time. The many experiences caused him to ponder questions on the fundamental nature of life. Some of these questions were implicit in the story of the banana plant in Robert's garden in Shrewsbury.

'My father is the largest man I've ever known,' Charles once said. Robert Darwin was called 'the Doctor' and had grown up in the shadow of his own larger-than-life father, the liberal thinker Erasmus Darwin, who had been one of a group of friends in the industrial north midlands of England called the Lunar Society. Reacting to the industrialization and social changes in Britain's heartland, this gathering of businessmen, engineers, scientists, artists, the potter Josiah Wedgwood and Erasmus Darwin met each month by the light of the full moon for a good night of talk in a Birmingham public house. If the verse that Erasmus is said to have composed for these occasions is anything to go by, they were jolly evenings:

> O mortal man that liv'st by bread,
> What makes thy nose to look so red?
> 'Tis Burton Ale, so strong and stale,
> That keeps my nose from growing pale.

After centuries of high authority, an establishment comprising landowners, church and state, these new liberals saw that it was time for change. The Lunar men were moved by radical reactions to the social upheavals that were upsetting the stable church and landowner control. This diversity led to a comfortable eccentricity in the Darwin family, which stimulated many of Erasmus's children to challenge the big ideas about the meaning of life.

For Erasmus and his friends, man was the pinnacle of all nature, as was floridly expressed in his poem, *Temple of Nature*:

> Imperious man, who rules the bestial crowd,
> Of language, reason, and reflection proud,
> With brow erect who scorns this earthly sod
> And styles himself the image of his God.

One of Erasmus's main themes was that if the machines of industry could shift the whole style of life in England, then nature itself was driven by some comparable force. He desperately looked for clues about how it worked, seeking out whether living things were fixed or flexible, targeted or random. The early nineteenth century was alive with the excitement of new ideas from the Enlightenment, not least of which were challenges to God's grand designs. Erasmus asked whether all species were linked together in some way, somehow related by a hierarchical progression. He went on to speculate that such transformation of species could even follow the same pattern for all species, placing humans at the pinnacle of this process.

Charles had read his grandfather's argument in the two books he had written for a medical readership: *Zoonomia* published in 1796 and *Phytologia* in 1800. In these works he argued that all organisms owe their existence to a process of advancement through long periods of time, governed by the laws of nature rather than something more divine. But fifty years later, Charles found both books far too theoretical and they left him 'much disappointed, the proportion of speculation being so large to the facts given'. The sense of excitement about experimentation was too great for Charles's generation to take their grandparents very seriously. The new scientific methodology was becoming the new spirit of power.

At the same time as this new way of looking at the world was coming into fashion, Erasmus and the Lunar Society celebrated the glorious effects of man's genius and industry on society, and how both were moving away from the wild vagaries of nature. But Erasmus genuinely loved animals and plants and bought several acres of land just outside Litchfield, where he created a botanical garden. In the 1780s, through the inspiration of the explorer Joseph Banks, he led a group that translated the books outlining Linnaeus's new system of naming animals and plants, *Systema Vegetabilium* and the *Genera Plantarum*, published twenty years before. Later, in 1792, he published a long poem that he called *The Economy of Vegetation* and its many admirers knew as *The Botanic Garden*:

> Led by the Sage, Lo! Britain's sons shall guide
> Huge sea-balloons beneath the tossing tide;
> The diving castles, roof'd with spheric glass,
> Ribb'd with strong oak, and barr'd with bolts of brass.

And then the cry:

> 'Let there be light!' proclaim'd the Almighty Lord,
> Astonish'd Chaos heard the potent word;

Thirty years after their publication in *Zoonomia* the young Charles was aware that these views were still causing disquiet in intellectual circles. Erasmus's published poems were still talked about. They had impressed Wordsworth, 'frankly nauseated' Coleridge, and been considered 'most delicious' by Horace Walpole. Such strong and opposing reactions had appealed to Erasmus's cheeky liberal attitude and fed into his hectic and hearty attitude to life.

When he was a young man Erasmus had been footloose and fancy free. Now, Charles learnt that after his grandmother's early death in 1770 his grandfather had attempted to reform himself, yet 'it proved easier to give up Bacchus than the charms of Venus' and he continued to have many children by various women. Even more shocking to his grandson were harsh realizations about the next generation, Charles's own father and uncles. Of these three sons by Erasmus's first wife, the youngest is thought to have committed suicide by throwing himself into the river at the bottom of his garden. The eldest also died mysteriously when he was a medical student, just after performing a post-mortem examination.

After these catastrophic events, the middle son, Robert, already an anxious young man, was concerned to retain some stability in his life. This determination helped him become a successful doctor and he increased his wealth as a shrewd investor. Robert married Susannah Wedgwood and became well known in Shrewsbury as an owner of property and as a supporter of many local business projects. Robert and Susannah's two sons inherited the family Christian names, Erasmus (Ras) and Charles, and both knew their father as the kindest of men.

But bad fortune continued to haunt the Darwin family: Susannah died when Charles was eight years old. With such a background, a bullying grandfather, two uncles dying tragically and a mother dying when he was so young, Charles was laden with guilt. The traumas of childhood would culminate in the young man's high state of anxiety, distrust of male institutions, a very close early relationship with his sister Caroline and another bond formed later with his wife Emma. When his own daughters Mary and Annie died very young his pain went especially deep. He had a constant need for the kind of reassurance that was provided by his wife Emma and the remaining children and for the excitement and intellectual curiosity of his scientific experiments.

Learning in Edinburgh and Cambridge

At medical school in Edinburgh during 1825 and 1826 Charles Darwin first came across the fundamental issues that had influenced his grandfather's ideas about the progression of species. The school also gave him the formal and straightforward up-to-date facts of anatomy and physiology that the eager Charles absorbed, but at the same time found wanting. He said the lectures on human anatomy were as dull as the lecturer himself, John Barclay, and he hated dissections. The lessons on medicine and biology lacked excitement and any deep sense of passion. There was no attempt to consider the meaning of life, nothing to hold on to and help search for some scientific explanation of why the whole of nature existed and how it worked.

To learn about these wondrous ways of nature Darwin turned to Robert Grant, sixteen years his senior and now back in Scotland after studying natural history at the Musée National d'Histoire Naturelle in Paris where he had been introduced to the latest ideas. In contrast to the boring lectures at Barclay's school Charles was fascinated by his trips with Grant around the nearby Scottish coast. He soon fell in love with the beauty and balance of the natural history, the sure patterns and styles of living together that he saw in nature, such as the barnacles and anemones of the Firth of Forth. He had yearned for this certainty at home and had never experienced it. Grant's explanations of how they all related to one another, however, was more difficult to accept. There was talk of what his grandfather had thought about the way different species might be related by transmutation, some kind of ancestral linkage. Darwin was not so convinced.

By the time he left Edinburgh in 1827 at the age of nineteen, Charles was deeply committed to the study of nature. Nonetheless, he was also sure that he didn't want to follow his father and grandfather as a doctor. The work was too close to the tragedies in his family and he had no desire to spend his life close to morbidity. He wished to explore the wilds of nature and find new living forms. He went to Cambridge where his father hoped he would train for the Church. But once there Charles realized that he was able to keep up with learning about biology. In particular he struck up a friendship with John Henslow, professor and founder of the University Botanic Garden, with whom he would shoot birds and collect beetles.

In his autobiography Darwin admitted that this time spent in Cambridge did not gain his father's approval. In one letter Robert admonished his son: 'You care for nothing but shooting, dogs, and rat-catching, and you will be a disgrace to yourself and all your family.' Nevertheless, Charles listened to those around him more than his father knew. In the 1820s Cambridge academics were beginning to understand the importance of unifying explanations that come from scientific laws. One hundred and thirty years after Newton had opened up the physical sciences Cambridge philosophers like William Whewell, the polymath who first coined the word 'scientist', and astronomer John Herschel were looking to science for an explanation of the meaning of life. The Reverend Adam Sedgwick, professor of geology, summarized the mood: 'The secure roads of honest induction make science a religion giving truth.' But at that stage in his career the physically active Charles was more interested in exploring nature and gathering evidence than in trying to interpret it.

By the end of his time at Cambridge in 1831 Charles was no nearer entering the Church than when he arrived. To end his course there he accompanied Sedgwick on a field trip to north Wales. They mapped the ancient rock formations where they found plenty of good fossils. With these they began to formulate original thoughts about the evolution of the landscape through geological time. Darwin's new relationship with this influential man turned out to be very timely and, together with Henslow's opinion, good reports were passed on to the Admiralty who were looking for a strong and intelligent scientist to join an important surveying voyage around South America and Australia. Henslow's advice was very persuasive: 'There never was a finer chance for a man of zeal and spirit'; the position was 'more as a companion than a mere collector. I think you are the very man they are in search of.'

Charles's father was horrified when he first heard of his son's prospective recruitment on the *Beagle* – and for the same reasons Charles was desperate to go. Robert Darwin wrote trying to dissuade Charles with as many arguments as he could think of, a list heavily weighted towards the cautious doctor's values. In response a frustrated Charles had the brainwave of showing the list to his uncle, Josiah Wedgwood, who reported back to his brother with another list of arguments in favour and the comment that the young adventurer would be 'deuced clever to spend more than his allowance on board the *Beagle*'.

It was just the right witty intervention that was needed to encourage the doctor to lift his concerns, which he did with his usual generosity: 'they tell me you are very clever'. He paid for his son's costs and living expenses as well as giving him £60 a year to cover the salary for a servant and collecting assistant, Syms Covington, who stayed with Darwin until he got married. The only other objection was from the 26-year-old captain of the voyage, Robert FitzRoy, who was very concerned about the strange shape of his recruit's unusually large nose.

Reading and talking on the *Beagle*

The autumn of 1831 was a hectic time for Charles. He met Captain FitzRoy for the first time and they went off to see the ship, which was being prepared at Devonport. In order to get to know one another they took the three-day trip by steam packet from Greenwich, around the Kent peninsula and along the south coast. The two gentlemen got along famously though it soon became clear to the captain that Darwin didn't respect the value of money very much and to Charles that FitzRoy, being a high Tory, didn't value human rights at all. But they did have many friends in common and saw their interests in science as something important that bound them together.

One of the first things that Captain FitzRoy did for his new cabin mate, or 'gentleman companion', was to give him a copy of the first volume of Charles Lyell's *Principles of Geology*. The first part had been published in 1830, almost two years before. Lyell was one of the few professional academics to work outside Oxford or Cambridge and had become highly influential within scientific circles. The book became serious reading for the young Darwin on the expedition. The ideas it set out were also to become important topics for debate between the two shipmates for the next three years and Lyell's argument for slow and gradual change through geological time became a central theme in Darwin's writings. He later would report that 'without the principles there would have been no *Origin*'.

Lyell was a powerful authority in British science and he and his family were to become great friends of the Darwins both in London and later in Kent. Lyell had taken the ideas of slow and uniform change in geology from two other men of early Scottish science, John Playfair and James Hutton. With their contemporary, Englishman William Smith, these men were the founders of geology, and were quick to calculate that the earth's history lasted many millions of years. Their studies encouraged them to think that there had been gradual change in nature. This was in stark contrast to views on the same matter which were beginning to emerge from Paris. It was a division of opinion that reinforced the political atmosphere of the early nineteenth century in which the aftershock of revolution and the Napoleonic Wars could still be felt.

At the time, the French were at the cutting edge of speculation on the theory of life. Influenced by the chemist Lavoisier, who had results from new experiments that showed that spontaneous generation was not how life began, Jean-Baptiste Lamarck had been looking for a single theory to explain life, not by sudden transmutation from one species to another but by some chemical-like process. This thesis was revolutionary for it suggested that all natural entities were in a state of flux. As in a chemical reaction one species would be transformed into another 'given time and favourable circumstances' and through sufficiently long periods of time these changes could explain all biodiversity. It was an altogether more serious and professional outline of what Erasmus Darwin had been saying at the Lunar Society yet it still had its detractors in England in 1837, and entomologist Reverend William Kirby wrote in one of the *Bridgewater Treatises* with a typical Victorian judgement of the Frenchman: 'In words, he admits the being of a God. He employs his intellect to prove he had nothing to do with Creation.'

Lamarck had been a soldier, then studied botany and published the *Flore Française* in 1778. He was a more serious man than Erasmus Darwin and worked as an assistant botanist at the Jardin du Roi in Paris until 1793, the year Marie Antoinette and Louis XVI were guillotined, when he became professor of 'insects and worms' at the new Musée National d'Histoire Naturelle. Lamarck's best-known work, *Philosophie zoologique*, was published in 1809, the year of Charles Darwin's birth, and celebrated an inherent 'tendency of life to improve itself in the course of time'. It was an argument for the gradual ascent along the scale of plant and animal beings, an upward progression from simplicity towards complexity causing one species to be replaced by another. He had seen such sequences in the fossil shells from the rocks around Paris, where some species appeared as similar to modern molluscs from the tropics that had been collected and displayed in the museum.

Coinciding as it did with Napoleon's successes and the creation of the French empire, Lamarck's ideas of transmutation became associated by many English commentators with revolution and Paris street riots. Together in the captain's cabin of the *Beagle*, FitzRoy and Darwin fuelled their aversion to Lamarck's notions with jingoistic anger at French arrogance and sedition. Most of all, this resentment was brewed in the fear of atheism. Lamarck showed a picture of life that did not need a god. Man was alone without an ordained hierarchy.

At the time there was suspicion that atheism was spreading and news had reached FitzRoy that a popular left wing group was selling illegal penny newspapers in the London slums praising atheism and the expectation of human progress for everybody. The ringleader was secularist George Holyoake who was imprisoned more than once for this blasphemy. Another man of the times with similar views was the utilitarian social reformer Jeremy Bentham, who in 1826 founded the godless institution of Gower Street later named University College London. Bentham believed in 'progress' and proselytized the idea that it was man who was responsible for delivering the world towards perfection.

Some of these views had been published in the 1809 volume by Lamarck, and Darwin had learnt of them when he was a student in Edinburgh. Yet the new ideas which were now being developed in Paris were in stark contrast to those of Lamarck. Georges Cuvier was a young God-fearing man born forty-five years before Darwin. He believed that species were stable and were created at different times so there was no need for them to evolve or change in any way. They were fixed entities and grew according to some grand design of the Creator. To provide sufficient space for new species Cuvier approved of extinction, whether it was caused by the deluge or by humans, and which usefully explained how fossil forms had gone out of existence.

The new First Consul, Napoleon Bonaparte, had appointed Cuvier as professor of natural history at the prestigious College de France. Cuvier had no time for Lamarck's earlier ideas of a fluid and slow transition from one form to another. Such inheritance by acquired characters was incompatible with the sudden changes that the Creator's restocking of new designs required.

Darwin's friend Grant had been in Paris as Cuvier and his more worldly colleague Etienne Geoffroy St Hilaire were beginning their notorious argument about how to interpret the life sciences. It reached a peak in 1830 at l'Académie des Sciences, in a debate about whether biological form follows the standard designs laid down by the Creator or whether they are more flexible adapting to what goes on outside in the environment. The first theory was rigid and measurable, the other, in the words of Geoffroy, like 'a poet trying to sing the grandeur of the universe in another form'. That was how Geoffroy spoke of it, under the influence of his friend, the German poet Johann Goethe, when he was asked to summarize the debate a year later. These were fundamental differences that came between the two Frenchmen, one representing a formalist view and the other a wilder and more outward functionalist one.

Grant had known of some of these exciting arguments before he had arrived in Paris. Cuvier's *Theory of the Earth* had been translated in 1813 and the book explained the disappearance of genera and species by extinction in various kinds of catastrophe, the 'last revolution' being the Ice Age. The *Theory* explained how alternations of floods and droughts could result in the drowning of terrestrial organisms and the suffocation of marine ones. It was catastrophes like these that decided the shape of life. There was no need therefore for ideas of evolution – change was driven from within the organisms themselves. Cuvier worked next to his ageing colleagues at the Paris museum who were devoted to Lamarck and who still had little time for history and fossils. Instead they were sure that the living forms of what some thought to be extinct species had just not been found alive yet. They were what Lamarck had called 'living fossils'.

On the other hand Cuvier suggested that different parts of animal and plant embryos developed from different designs laid down by the Creator. It was how different species were formed and how closely related species had so many features in common. The Creator designed a particular series of segments for each individual within the same species, varying only a few of the segments when only small changes were required. When a new design was needed after a catastrophe the new species was a product of special creation and immediately replaced the earlier extinct ones. These were the topical thoughts that stimulated the often fierce arguments between Captain FitzRoy and Darwin in the cabin late into the night.

Biodiversity and geology from the *Beagle*

HMS *Beagle* had been commissioned to survey the islands and coastal channels around South America and the Pacific for trading routes and the Royal Navy. An ancillary purpose was to collect specimens of natural history, crate them back to Henslow in Cambridge, and make observations of the geology and geographical distributions of the flora and fauna. About two-thirds of Darwin's time was spent on land and for the rest he was one of the seventy-four men on the 90-foot (27.4-metre), ten-gun brig. The 22-year-old slept in the larger poop cabin with the assistant surveyor and midshipman, both still in their teens, and he also had access to the captain's own cabin during the day.

After three weeks at sea the *Beagle* made its first call at St Jago on the Cape Verde Islands, 300 miles (483 kilometres) off the African coast. It was the first subtropical vegetation that Darwin had seen and he was thrilled that those plants that he had only known from pictures had such warmth and fresh colour in real life. The palms moved with a brushing in the breeze, the giant baobab trees gave a ghostly sense of the past while wild cats and kingfishers darted back and forth. Away from the forest a black volcanic landscape showed raised beaches stepping up from the sea with layers of redundant and now dead coral and seashells. Henslow and Lyell would be pleased with these first pickings and Darwin himself couldn't stop noticing how similar much of the nature was to that of the descriptions he had read of mainland Africa.

It was different, for sure, from what they were to see on the other side of the ocean. At Rio, Darwin and his cabin boy and expedition assistant, Syms Covington, collected butterflies, fireflies, worms, frogs, ants and spiders. They saw vast tropical rain forests and got used to the warm and humid climate. At Montevideo they found other birds and snakes, pampus and more raised beaches, and the blind rodents and amphibians that reminded Darwin of Lamarck's arguments. These unexpected specimens gave subject for discussion in the captain's cabin. Darwin's *Journal* of the voyage records the arguments he had with FitzRoy, usually lively and friendly, on topics such as the subterranean vertebrates that seemed to have no need for sight and were therefore blind.

Yet in time the disputes became increasingly hostile. When they turned to discuss the origin of the plains along the coast of Patagonia the arguments even became nasty. The different viewpoints became more entrenched as the voyage continued. 'I have harboured a viper in my bosom,' FitzRoy wrote later, when he read Darwin's dismissal of his favoured catastrophic formation of the coastal plains of Argentina. Instead of the deluge Darwin proposed the gradual rise of the pampas, still influenced by Lyell's ever longer timescale for the history of the earth.

At least the debates with the captain made Darwin clearly aware of the kind of opposition he would have to face back home when explaining his results and observations. Some of his findings from Buenos Aires in October 1833 and from Patagonia the next year were of fossil bones belonging to large horses and mammoth-like creatures and were hard to fit into any theory. Many of these fossil mammals were of unknown age and so it was hard to know whether they supported slow or sudden geological changes. There were also skeletons of unknown species of monkey, anteater, armadillo and sloth. Even when the specimens were unpacked in Cambridge their significance wasn't clear, adding to the excitement that Darwin's explorations were causing.

Other experiences were to have a huge impact on Darwin's thinking but they were so difficult to understand fully that he was very careful about what he said. Like Hamlet he could not make up his mind. One experience was to challenge the slow and gradual outlook he had learnt to expect from geological processes. While they were docked at Valdivia in southern Chile in February 1835 there was an unexpected event that the whole crew felt, heard and saw. The earthquake and associated volcanic activity were not the kind of events that readers found in Lyell's slow-moving *Principles of Geology*. Covington described the event vividly: it 'came up of a sudden' and was like being 'in a gentle seaway'. In his *Journal* Darwin compared it to skating on thin ice:

> An earthquake like this at once destroys the oldest associations; the world, the very emblem of all that is solid, moves beneath our feet like the crust over a fluid . . . In the forest, a breeze moved the trees, I felt the earth tremble, but saw no consequence from it.

The earthquake was followed by a giant tsunami. Similar events had occurred in the same region in 1751 and 1822 so the population had learnt to run up into the hills. At Valparaiso:

> all the houses were built of wood, none actually fell & but few were injured. Everyone expected to see the Church a heap of ruins. The houses were shaken violently & creaked much, the nails being partially drawn. I feel sure it is these accompaniments & the horror pictured in the faces of all the inhabitants, which communicates the dread that everyone feels who has thus seen as well as felt an earthquake.

An old woman, who was on the beach at low tide told Darwin that 'the water rose quickly but not in big waves to the high water mark & as quickly returned to its proper level'. FitzRoy reported more trauma at Concepción: 'Roaring as it dashed against every obstacle with irresistible force, it rushed – destroying and overwhelming – along the shore. Earth and water trembled: and exhaustion appeared to follow these mighty efforts.'

A few months later the *Beagle* sighted the Galapágos Islands 600 miles (966 kilometres) west of mainland Ecuador. Charles had written home to say that 'I look forward to the Galapágos with more interest than any other part of the voyage. They abound with Volcanoes & I should hope contain Tertiary strata.' Sure enough, each island had originated from volcanoes growing and erupting separately about ten million years ago. Consequently, the first animals and plants to colonize the islands had to travel across the sea and each community continued in isolation. As well as the finches and other unique genera of lizards and tortoises almost half of the species of flowering plants were restricted to the islands. Most of these different species were confined to separate islands and were closely related in small groups that comprised different genera. The visit affected Darwin deeply as a 'little world within itself'. There he felt that 'we seem to be brought somewhere near to the great fact – that mystery of mysteries – the first appearance of new beings on this earth.'

The rest of the cruise across the Pacific to Australasia and then around the Cape of Good Hope added more crates of specimens to amaze the specialists back home and provided more observations of the global patterns of animal and plant distribution. In Australia, the kangaroos and possums seemed to be everywhere and the colourful flowering shrubs of *Banksia* and *Protea* loved the dry slopes of many southern hemisphere habitats. It was still a mystery why these great groups were absent from the northern hemisphere and how they related to those that were. And horses, holly bushes, oak trees and other groups well known from the north showed up unexpectedly in the southern hemisphere, far away from where they were most familiar.

The scale and force of the five-year voyage would be immense. It is no wonder that it took Darwin several years to gather his thoughts into one coherent argument about the many threads of different magnitude and pace. But one thing was clear early on: the boxes and crates that he had gathered contained clues to a new world of ideas. Darwin's experiences would force him to reassess everything he had so far accepted.

A bachelor in London

In October 1836 Charles Darwin was back in a Europe gripped with apprehension, trying to understand the state of the world. It seemed as if the volatile political climate was moving in exactly the same way as the current debates on the transmutation of species. Would change come gradually or by revolution and catastrophe? Nobody seemed to have the answer. Some still hoped for gradual change while most demanded a radical shift and were willing to fight for the right to vote and the abolition of slavery.

Darwin spent the next six years of his life in London, living the style of a young gentleman. He moved into rooms in his brother Ras's house in Great Marlborough Street. His time was spent socializing and rushing through his allowance from his father. The return of his brother gave Ras's supper party guests new topics to talk about and there was no shortage of intellectual friends to introduce to the recently returned explorer. They included Whig lawyer and politician Thomas Macaulay, Utilitarian leader John Stuart Mill, Charles Babbage the inventor of the calculating machine, and Sir Charles Lyell. There was also no shortage of attractive young women such as Fanny Wedgwood and Jane Carlyle, the fashionable Chelsea hostess who was unhappily married to the polymath Thomas Carlyle. The unmarried writer and campaigner for women's rights Harriet Martineau was another of Ras's visitors, and she was there frequently enough to inspire gossip. The Darwins' soirées bubbled with the fashionable reasoning of the new Utilitarian thinkers such as Mill and Jeremy Bentham.

This was an intellectual group at the centre of the social changes that were going on all over Europe. The discussions were about the meaning of life, how science promotes a different kind of meaning and how this might put self-interest at the centre of morality. For Charles, the idea of the individual moving forward with such a force was the obvious way of explaining change. His group of friends agreed that when there were sufficient numbers of these individuals changing in the same way, then that would benefit the whole system. It was a process that could be passed from one generation to another by family tradition and education, just as Lamarck had advocated. Yet Darwin was looking for a more structural explanation of change, a system that caused and transmitted the changes inside the organisms from one generation to another. In those Victorian drawing-rooms the new spirit of capitalism looked with favour at the individual. With his main interest centring on natural history, Darwin was inevitably going to extend this view to groups of individuals in other species.

At one of the dinner parties in 1838 Harriet Martineau recommended Charles read Thomas Robert Malthus's *Essay on the Principle of Population*, written in 1798. The book suggested there was a natural tendency for populations to increase faster than the means of subsistence. As an economist, Malthus offered elaborate proofs from human history showing that a population declined once a peak was reached. Full of observations and thoughts from his voyage Darwin immediately connected Malthus's ideas to other species. He wondered if the sizes of populations were limited by the availability of food and space.

Malthus had revealed that the greatest influences on population growth come from outside the group. Darwin's great breakthrough was to realize that natural selection was a way out of this trap. It would be the strongest that would survive, while the weakest would perish. Not for him Jeremy Bentham's phrase 'the greatest good for the greatest number'. With natural selection there had to be death for survival, for evolution to occur there had to be extinction, with success there had to be failure.

Darwin had heard talk of this kind of conflict for the first time at a staging-post on the tip of South America four years earlier. In South America he had picked up a parcel sent at Christmas by his three sisters which contained a copy of Harriet Martineau's *Poor Laws and Paupers Illustrated* (1834), the author's philanthropic argument urging sexual restraint to reduce the population and cease starvation. Darwin's initial reaction was not entirely sympathetic: 'Erasmus knows her & is a very great admirer & every body reads her little books & if you have a dull hour you can, and then throw them overboard, that they may not take up your precious room.' Away from the intellectual stimulation of Great Marlborough Street Darwin had missed an important clue of Malthus's evidence. Four years later, and in the heart of the metropolis, Darwin understood the significance of how a rapid change can cause a struggle for existence.

One of the most tempestuous relationships that Darwin established in these bachelor times was with Richard Owen, the ambitious professor of anatomy at the Royal College of Surgeons in London. At the start both men got on well together, partly because they were introduced in glowing terms by the powerful Charles Lyell, and also because of the good things each had heard said about the other. To seal their initial friendship Owen had jumped at Darwin's offer to be the first to examine the crates of fossil mammal skeletons from South America. This proved to be a successful exercise and the monograph *Fossil Mammalia* that Owen published two years later, in 1838, became a classic demonstration of how similar extinct species can be to their living relatives. There were descriptions of the 60-million-year-old small Dawn Horses and Pseudo-Horses, the 30-million-year-old Pseudo-Mastodont looking like an elephant, and others like rabbits and hippos. These fossil species were all closely related to other mammals found close by in geological time as well as in geographical space. The work boosted both men's reputations, Owen for his argument that they were good examples of close modular design and Darwin for having found the fossils in the first place.

It was on the crest of this first success that Darwin first contemplated starting a family. Charles was approaching his thirtieth birthday and the question of having children of his own had to be answered. He made another list, this time of the advantages and disadvantages of matrimony. The criteria read today as explicit signs of the times, a commentary of what he intimately held dear alongside items that few of us would like to admit even internally. On one side he listed 'children (if it please God), constant companion, home, music'. On the other were 'freedom to go where one liked, conversation of clever men at clubs, not forced to visit relatives, the expense and anxiety of children, perhaps quarrelling, loss of time'.

In January 1839 Charles Darwin and his cousin Emma Wedgwood were married quietly at Maer in Shropshire. They moved into Macaw Cottage in Upper Gower Street, in the middle of Bloomsbury. It was busy, noisy and dirty, though for Charles it was convenient: next door to Bentham's new university, just round the corner from Owen's office at the British Museum, and not far from the newly opened Euston Station. For Emma everything was very different to the country life she was used to and it was soon clear there was not much for her to enjoy.

Moving to Downe

Emma always suspected that Charles wanted to do better than his well-known grandfather Erasmus. At her family home in

Staffordshire there was a different style of living altogether, more relaxed and conducive to thinking, and she was convinced that this was the environment in which her new husband would thrive. In May 1842, Charles went with Emma and their two children to Maer. Pleased to leave the turmoil of London, Darwin was able to concentrate on his ideas on nature. It was the ideal place for him to consider the ideas that had emerged from his voyages and his most recent reading. Since his return from South America his mind had been churning up many thoughts about species progression and now he wanted to put them into some kind of order. In the Maer garden that spring he wrote hard for a week and drew up a 35-page draft. It was the first written account of the adaptation and origin of species by natural selection.

Although the draft essay was lost for a time, one version does survive and shows that by that summer Darwin had worked out the main features of his theory of evolution. From Malthus he had worked out the central theme of competition. It also reveals that he had learnt a lot about the process of adaptation from his reading of magazines for amateurs interested in animal and plant breeding. Further revelations in the essay came from his voyage on the *Beagle.* His suspicions had grown that similar anatomical structures came from the same parental stock and this had got him thinking about a theory of common descent. It led on in the same draft essay to the idea that whole communities are organized through a survival strategy, driven somehow by sexuality. Yet the proposal raised one major hindrance: his ideas were still hazy and he knew that without sufficient scientific evidence they would not stand up to Royal Society scrutiny. They would be ridiculed and lost. If only he had the facilities to begin some of his own experiments.

The summer of that year was particularly hot in London, to which they had returned in July, making life at Macaw Cottage more uncomfortable than normal. Added to that, a month-long national strike, protests against the introduction of workhouses for the poor, brought crowds on to the streets. Several Lancashire cotton mills were closed by the rioters who tried to stop troop movements on the new railway going north from Euston. Emma and Charles's house was so close to the station that they were uneasy with the sight and sound of the troops that kept marching past, sometimes with fixed bayonets, clearing a way through the mob so that more troops could board the trains to Manchester.

More mundane difficulties with the London house also appeared – the garden was inadequate for a young family: 'dingy grass in the garden, a strip as wide as the house and thirty yards long'. The young couple were torn between living in the metropolis or in a more rural idyll. In 1839 Charles had written to his cousin William Fox:

> We are living a life of extreme quietness: and if one is quiet in London, there is nothing like its quietness – there is a grandeur about its smoky fogs, and the dull distant sounds of cabs and coaches; in fact you may perceive I am becoming a thorough-paced Cockney, and I glory in thoughts that I shall be here for the next six months.

But his diary was littered with entries concerning ill health that told a different story.

With two small children and the servants, Macaw Cottage was too small for the growing family. Later in July 1842 Charles went to stay with his father and sisters in Shropshire and wrote to Emma, 'My father seems to like having me here; he and the girls are very merry all day long. I have partly talked over with the Doctor about my buying a house without living in the neighbourhood.' Dr Robert Darwin once again followed the pattern of opposing his son's plans at first and then generously supporting them. He promised Charles enough money to buy a house and gave his blessing to the family finding a new home in the country.

In mid-September the Darwins moved to Downe. Emma was eight months pregnant, and happy in the knowledge that the new home would be suitable for a growing family. Here too Charles would find it easier to carry on with his writing. But Mary, their third child, born nine days after they moved to Down House, died in October. Added to the earlier deaths that he had experienced, this further grief haunted Charles with nightmares of guilt and fear, which drove him to bury himself in his studies. Darwin's description of the place just after this tragedy says a lot about his mood late that year:

> In 1842 it was dull and unattractive enough; a square brick building of three storeys, covered with shabby whitewash and hanging tiles. The garden had none of the shrubberies or walls that now give shelter; it was overlooked from the lane and was open, bleak and desolate.

The house had been built in 1778 on the site of a farmhouse dating from more than a hundred years before. The Darwin family made several alterations from 1843 to 1878, adding a bow window and veranda at the back and a new hall and study. In 1843, to stop the intrusion of people walking down the lane they lowered the road 2 feet (0.6 metres) and built a flint wall. In 1842 they employed Brodie, a Scottish nurse, and two years later they needed a governess. Joseph Parslow, who ran the house and served as the butler, had joined the family in Gower Street and stayed until he was pensioned off in 1875 when he was given a cottage on the estate.

In the first few months at Downe, Charles could not leave London behind completely. He wanted to share the ideas of his recent essay with his friends and colleagues by putting them into print as soon as they were presentable. At Downe, he could concentrate in isolation. Yet as he thought through his notions he became aware of some of the religious and philosophical implications of what he was thinking. The more secrets he was unearthing through his observations, collections and experiments, the more new and difficult conclusions came to light.

It would soon become clear to all members of the household that much more was going on in the apparently quiet garden than in the normal gentleman's plot. At Down House Darwin discovered a place that might offer clues about transmutation, different from what was then being offered in museums and scientific societies. Away from the hubbub of the city, life in the country enabled Darwin to observe what went on in the meadows

Verandah 1872

New drawing room 1858

Front door 1858

New study 1877

Front door and porch 1877

Hall

Front door 1778

Old study

c.1778

Bow window 1843

Dining room (old drawing room)

Old dining room

Meat Store

Site of original farm house c.1650 kitchen

Pantry

Scullery

Back door till 1845

Back door after 1845

Window bricked in to avoid Window Tax (1695–1851)

Plan of Down House

and woodlands, experiment in the greenhouse and explore the hidden small dimensions inside the cells under his microscope. What first appeared to be an isolated and quiet Victorian retreat would turn out to be the proving ground for revolutionary change.

As well as giving his family the thought and care of a loving father, Charles also gave the same skilled attention to his beloved garden. Here he explored the first tests of his theories about plant and animal life. The experiments were designed to expose his ideas to the most rigid of scientific methods. Darwin had no intention of turning the 16-acre (6.47-hectare) estate of Down House into a place of study but it was exactly what happened. There, all on his own, his thoughts were able to build on his experiences more effectively than he had expected. Away from London society the quiet and reserved Charles became more involved with how to test these thoughts and he quickly turned the garden into a living laboratory.

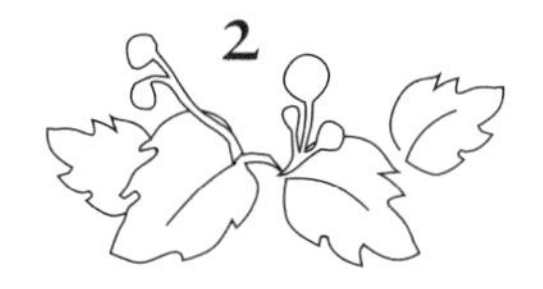

2

A New Garden at Down House

Designing at Down

The garden was about to come to life, not only with a scientific role but also with play and as a sanctuary for the growing Darwin family, their relatives and friends. The tired 16 acres (6.47-hectares) sprang into activity with new plans and projects, which came as a shock to the two resident gardeners. But the Darwins and the gardeners soon got used to one another and started to enjoy many years of fun and hard work.

William Brooke was known to the family as the 'gloomy gardener' for they only saw him laugh once – when Darwin tried to throw a boomerang round the lawn and it fell into one of the glass cucumber frames. The taciturn Brooke also looked after the cow and the pigs and helped the governor with his experiments, planting out the sundew and primulas. Henry Lettington was Brooke's son-in-law and he was the apprentice gardener. He taught the young Darwin boys how to make a whistle in spring and helped them tame their rabbits. The whistle came from hedge parsley which grew in profusion along the hedgerows and beside the footpaths every summer. The small white flowers shone like spraying fireworks up to the children's waists and Brooke showed them how to make a hole in the hollow stem and use a blade of grass as the reed.

The children didn't go into the yard by the Home Meadow, south of the house, for that was Brooke's territory, his place of work with the horses and the other animals. It was a busy place and kept the household in eggs, meat and milk. Fresh water came from the well at the side of the yard, not far from the back door into the scullery. Some of the children said it was 365 feet (111 metres) deep and the long rope that held the bucket was kept by the mulberry tree where the blackbird sang every morning.

Darwin built a small wooden shed close to the well and, for a while, he went there every day, took off his clothes and, from an overhead tank, poured ice-cold water all over his body, the daily douche prescribed as treatment for his continuing stomach problems. Sometimes this remedy was followed by being wrapped in a soaking wet sheet and sitting by a very hot oil lamp, to make him perspire, and finally plunging into a bath of cold water placed on the lawn. More to his liking, each treatment had to be followed with a walk around the garden. At first he commented that 'the treatment is wonderfully tonic' but his stomach cramps returned and continued to trouble him for the rest of his life.

Francis Darwin thought that it was during these early years of married life that his father suffered perhaps more from ill health than at any other time of his life. As early as 1840, when he was just thirty-one years old, Charles wrote to his cousin William Fox: 'I am grown a dull, old, spiritless dog to what I used to be. One gets stupider as one grows older I think.' In 1845 he wrote to his new friend Joseph Hooker, son of the Director of Kew Gardens:

> You are very kind about your enquiries about my health; I have nothing to say about it, being always much the same, some days better and some worse. I believe I have not had one whole day, or rather night, without my stomach having been greatly disordered, during the last three years, and most days great prostration of strength: thank you for your kindness; many of my friends, I believe, think me a hypochondriac.

One form of relief from these attacks came in another wooden structure that he had erected in this same part of the garden

– a hexagonal pigeon loft. It fitted in the yard by the yew trees and gave Brooke and Lettington plenty of work every day cleaning each compartment and portioning out the right amount of food. Soon the birds became an accepted part of the local fauna; Darwin was happy that they didn't interfere too much with the other animals and plants because the food that the Down House kitchen provided was more to their liking.

For the first few years, however, no one had anything very kind to say about the house itself. Darwin wrote to Fox saying that it was 'very ugly' and spent the rest of the letter describing the joys of the surrounding countryside. Their daughter Etty, who was born in the house in 1843, later recalled that,

> the house was square and unpretending, built of shabby bricks which were afterwards stuccoed, and with a slate roof. It was of moderate size when bought, but was gradually added to, and became in time capable of holding a large party.

The first of the additions began three years after they moved in when they extended the servants' quarters of the house and replanned the garden. Charles's brother Ras came up from London especially to help by 'making a new walk in the kitchen garden, and removing the mound under the yews on which we found the evergreens did badly and which was a great blemish in hiding part of the field and the Old Scotch Firs.'

Emma took charge of the new design; she also cared most for the selection and planting of the flowers and shrubs in the beds beside the lawns and the borders beside the walls and fences. 'The repose and coolness of it is delicious,' she wrote; she enjoyed the sunny faces of the children playing, the smells of the fruits and flowers, the burring sounds of the bees and the flywheel of the well rattling. Her granddaughter the artist Gwen Raverat remembered that:

> Uncle Horace was once heard to say in a surprised voice: 'No, I don't really like salvias very much, though they did grow at Down.' The implication, to us, would have been obvious. Of

course all the flowers that grew at Down were beautiful; and different from all other flowers. Everything there was different. And better.

As well as choosing the herbaceous plants, Emma made sure there were plenty of more durable evergreens to protect and serve as a background. Beside the pebble paths of the Kitchen Garden small box hedges marked the boundaries. Ivy and holly mixed into the larger hedges and the evergreen scotch firs, while yews shielded the busy yard from the drawing-room veranda. From the drawing-room the lawn stretched out to the first shrubbery. On the left of the lawn Home Meadow was separated by a long low fence. In the foreground six large rectangular flower beds were stocked with Emma's changing displays of herbaceous flowering plants, admired whatever the weather from the windows at the back of the house. In the middle there were foxgloves and phlox, lilies and larkspur. Along the sides were smaller plants such as gazania, portulaca, primrose and aubretia. In the middle was a bird-bath, then a sundial by the side and later Horace Darwin's staked stone to measure how much soil was moved by worms. Outside the schoolroom window there was an old mulberry tree where the fruit made dark blue marks on the paving in the summer months.

The lawn edges of the flower beds were always clipped straight, creating long lines of perspective to the row of lime trees at the far end of the garden. Dorothy Perkins ramblers hung from a metal rose arch and opposite, over the fence and the meadow, there were panoramic views across the Weald to more chalk hills in the distance.

Gwen Raverat told many stories about her relatives at Down House and the whole experience of visiting there as a child made a great impression. One treat was to gather nosegays, small bunches of sweet-smelling flowers, with Aunt Bessy:

down the long pebbled walk between the tall syringa and lilac bushes all wet with dew, to the kitchen garden, where the roses were imprisoned behind high box borders, near the empty greenhouses, where my grandfather had once worked. We took the wooden trug of flowers, which smelt sweeter than any other flowers in the world, back to the house, and arranged them in water on a green iron table, in the Old Study, where *The Origin of Species* had been written.

On the north side of the garden, over the high hedge and shrubbery, was the walled garden where normally a botanist would have laid out formal displays of plants. At Down House it was called the Kitchen Garden and it had the usual rows of cabbages and carrots and a whole range of other useful produce set out with love and care. The seedlings were protected from the winds by more box hedges around the side of each plot and a high wall was built of flints from the chalk and red bricks from a local clay pit.

Along this long wall the greenhouse faced south and on its other side there was the orchard. This was a place of fun and games of high suspense for the Darwin household competing for the best fruit. Here, Charles first took the challenge from his clergyman cousin William Fox as to who was to grow the biggest pears from the latest varieties. They competed with surprise fruit from grafting different specimens and they tried out crossing different breeds over several seasons. The size of each year's crop was measured and became a subject of regular family entertainment. In this way, experiments in the garden became an important part of daily life for the household.

Beside the orchard, Darwin later built a red-brick potting shed as a place for experiment, where varieties, breeding programmes and hybridizations were tested and recorded. In winter months he warmed himself beside the small fireplace while he contemplated the meanings of the measurements that he had taken each day from plants grown in special conditions. This rhythmic way of life helped him put together the difficult shapes that he was recognizing in his painfully complex biological puzzle about transmutation.

As he wandered through the garden Charles Darwin saw no great interest in order, God or social progress. His garden was nature itself and the landscape its home. At Down House the garden was to be for observation and experiment, for the family's leisure and for planting cabbages. Both Charles and Emma grew up with nature in their lives, a love that was rekindled when they first saw the garden at Down House. Together they first looked past the then neglected flower beds, along the distant garden path into the fields beyond. It all reminded them of their childhood homes, that strong affinity with roots.

In all his solitude along these pebbled paths, the dark image of a troubled man was as tragic as the happiness of the children playing on the lawn was calm. Charles was putting together the pieces of a puzzle that strained him inside, reviving painful memories – his missing mother, uncles mysteriously dying – and making him ever more conscious of his wife's devotion to a God he could not share. Were the arguments going to be strong enough to set aside these concerns? He feared his cold alternative meant that life on this planet was a complex and self-organized organic system with no hope of comfort from a Creator. The ghosts of his childhood, the experiments to test these speculations, the horrible possibility of confrontations with so many people, all added to his loneliness and an accumulating sense of fear.

At the end of the Sandwalk beyond Home Meadow they decided to plant a copse. In 1888 Darwin's son Francis reminisced that it

> was planted by my father with a variety of trees, such as hazel, alder, lime, hornbeam, birch, privet, and dogwood, and with a long line of hollies all down the exposed side. In earlier times he took a number of turns every day, and used to count them by means of a heap of flints, one of which he kicked out on the path each time he passed.

Francis and the other children had often watched their father on these walks, once when

> some young squirrels ran up his back and legs, while their mother barked at them from the tree. He always found birds'

> nests even up to the last years of his life, and we, as children, considered that he had a special genius in this direction. He used to tell us how, when he was creeping noiselessly along in the 'Big-Woods,' he came upon a fox asleep in the daytime, which was so much astonished that it took a good stare at him before it ran off. A Spitz dog which accompanied him showed no sign of excitement at the fox, and he used to end the story by wondering how the dog could have been so faint-hearted.

During their first two years at Downe, Charles wrote a set of notes about his observations of the local natural history and called the collection *The General Aspect*. It set the atmosphere of the local scene as well as providing some details of botanical and ornithological record. At the end of March 1844 he described:

> The first period of vegetation and the banks are clothed with pale blue violets to an extent I have never seen equalled and with Primroses. A few days later some of the copses were beautifully enlivened by *Ranunculus Auricomus*, wood anemones and a white *Stellaria*. Subsequently large areas were brilliantly blue with blue bells [sic]. The flowers here are very beautiful and the number of flowers, together with the darkness of the blue of the common little *Polygala* almost equalled it to an alpine Gentian. Different kinds of bushes in the hedgerows were entwined by travellers' joy and the tree bryonies.

Then in June:

> The clover fields are now of a most beautiful pink and from the number of Hive Bees frequenting them, the humming noise is quite extraordinary. Their humming is rather deeper than the humming overhead that has been continuous and loud during all these last hot days, over almost every field. The labourers here say it is made by 'air-bees' and one man seeing a wild bee in a flower, different from the kind, remarked that 'no doubt it is an air-bee'. This noise is considered as a sign

> of settled fair weather . . . There were large tracts of woodland that were cut about every ten years, some of which were very ancient. Larks abounded and their songs were most agreeable, nightingales were common.

The ordering of nature

Darwin's passion for observation and recording and interpreting what he saw were also perhaps an extension of the common pursuit of the country gentleman. His study at Down was a strange mixture of the two habits: brass microscopes, mahogany furniture and piles of specimens and papers, all within a middle-class domestic setting. A lot of English country houses still have their 'Ark', a tall oak or mahogany cabinet like a small wardrobe with rows of drawers and a sloping top shaped like a roof. Usually each drawer is numbered though sometimes they are named to show the wide range of natural objects they once contained. They were kept in the hall or under the stairs next to the umbrellas and coats. Later, smaller Arks became fashionable as gifts at Christmas and the trinkets inside became more varied than the specimens of natural history that the first ones contained.

The early Arks stored small specimens of rocks, fossils, animals and plants, ordered according to the owner's chosen sequence. Each cabinet contained an interpretation on how the Creator was thought to have made everything in nature, stable minerals and species, in a sequence that became more complex and led up to human perfection. The cabinets also tidied up the collections and the labels made clear where each specimen was located in the arrangement of things.

In 1765 a formal order was given to the increasingly fashionable idea of progression of species in a strange little book by a Swiss lawyer and naturalist called Charles Bonnet. *Contemplation de la nature* proposed a fixed chain of being, the *scala naturae*, a ladder rising without any break in continuity, with no resting platforms or punctuations in the smooth flow of self-development. For Bonnet animals and plants were to change along this line to greater perfection. To start, corals led to truffles, then mushrooms to lichens and plants. On the same ladder he then stepped up to insects, fish, birds and finally through quadrupeds to man.

In the late 1760s Carl Linnaeus used a much looser scheme to order the botanical garden at Uppsala where he worked as the university's curator of plants. His model was based on the world's four great rivers thought to have flowed through the biblical Garden of Eden. Each quarter was marked out to plant six major groups based on their shared sexual characters, features of the flowers' petals, stamens, stigmas and styles. He saw the garden as a museum of God's Creation. In his system of classification Linnaeus used only structural similarities, features that the plants shared. He deliberately ignored all the differences between plants and all other features not related to the structures that he had selected for identification.

The aim of ordering nature with a simple system of naming was well made and is still in use internationally as the main way of reference. The twenty-four sections in the Uppsala garden became an important platform for the many names Linnaeus invented. They have become the basis of a system that classifies the differences by rank, like styles of military personnel. These hierarchical ranks go from species up to genus, Family, Order, Phylum and finally Kingdom (of which there are now five – plants, animals, fungi, bacteria and very simple cells called Archaea).

Yet Linnaeus's scheme posed one problem. On one hand he could classify a myriad of species within the binomial system that gave two names to each plant, referring to its genus and species, such as *Homo sapiens*. Yet this system was created at a time of rapid imperial aspirations. Plant hunters then, like venture capitalists now, were making new discoveries across the world and bringing them back to Europe for naming and classifying – but this would cause complications for universal usage.

After his experiences exploring in South America, Darwin was fascinated by attempts to work out the global migrations and distribution of animal and plant species. The mammal fossils that he found in Chile were records of migration as well as evolution, responses to the unstable environment of which he and Covington had borne witness in 1835. Charles had worked out from this that as a species moved across these vast regions there were often changes in shape and structure, reflecting even subtle differences in the environment on the journey. He had followed the stories of many professional explorers and collectors in the late eighteenth to early nineteenth centuries who had described a lot of new species of animals and plants for the first time.

These new discoveries came from all over the world and were defined by a wide range of features and criteria. Many of the authors didn't know what had gone before or which species had been found elsewhere, while others relished the chance of fame by being the first to name a particular plant or animal. The questions of where one species started and another ended, let alone in which genus and Family group they should be classified, were usually not considered very carefully. The resulting chaos was almost as bad as having lots of specimens with no more than nick-names.

The first serious English explorer of biodiversity in the tropics had been Sir Hans Sloane, an avid collector and artist of natural history. Sloane was a wealthy London surgeon and in 1687 was persuaded by the Duke of Albermarle, Governor of Jamaica, to join him as his personal physician. As a doctor he provided the duke with quinine from Peruvian bark and milk chocolate for his stomach cramps. Once he told his patient to swallow a hundred millipedes each morning and plenty of crabs' eyes later in the day. Jamaica gave Sloane plenty of time to amass a big collection of mammals, birds, fish and insects. But life there was not easy in comparison to what London offered him and he hated the climate:

> The heat and rain are excessive. The parts not inhabited . . . are often full of Serpents and other venomous Creatures. The same places remote from Settlements are very often full of runaway Negros, who lye in Ambush to kill the Whites who come within their reach.

He returned to London with crates packed with living and preserved specimens. Like many Europeans at the time, Sloane now set about trying to reconstruct alien plants and animals back in England. The new practices of plant and animal breeding had led to the popular cultures of horticulture in the English garden and animal husbandry on the farms and in the homes. Sloane had just purchased the Manor of Chelsea from Charles Cheyne and was able to plant his specimens there in the Physic Garden that had just been opened by the Society of Apothecaries. He leased them the 4-acre (1.6-hectare) site for £5 a year in perpetuity.

Almost a century later, from 1768 to 1771, another well-known English botanist, Joseph Banks, ventured beyond the tropics on Captain James Cook's voyage to Australia and Easter Island in *Endeavour.* Banks returned to England with hundreds of specimens of plants that he, also, had planted in the Chelsea Physic Garden, and donated the preserved specimens to the British Museum. He then persuaded George III to establish Kew Gardens as a botanical research centre for the nation. He also turned his mansion in Soho Square into a herbarium for his growing collection, importing plants and their produce from the growing British empire. To look after his plant collection he hired Linnaeus's brightest student from Sweden, Daniel Solander, and the great new collection of plants from all over the world was organized according to the Linnaeus scheme for its first big test.

As the gardens at Kew were being established, a friend of Banks with a strong reputation as a surgeon was becoming interested in how so many species of animals and plants had come about. It was not only their classification that was fascinating but their evolution as well. As a teenager in the 1740s John Hunter had learnt human anatomy by helping his doctor brother perform operations and make dissections. He also began a collection of stuffed mammals and pickled organs which would become a lifetime's task to maintain. Some are still on display in the Hunterian Museum of the Royal College of Surgeons, but none of the specimens appear as shocking as did the stuffed giraffe in the entrance hall of his Jermyn Street house. This independent-minded pioneer, today recognized as the founder of scientific anatomy, had practised his techniques on vagabonds and aristocrats and he later became surgeon to George III and William Pitt. His collection has become as symbolic for zoologists as Linnaeus's herbarium is for botanists, though Hunter's ideas about relationships between species were much more creative.

For the 1780s Hunter's experiments were original and ingenious. He opened chick eggs to display different stages of development in the embryos. He dissected the lymphatic system in sheep and confirmed its importance in draining waste products. His friendship with other naturalists like Joseph Banks gained him access to the Royal Menagerie at the Tower of London where he obtained a range of exotic dead animals for dissection. One such was the greater siren from North Carolina, an aquatic salamander with no external sign of hind limbs and only weak forelimbs. Hunter argued that these creatures had just two legs while the other two limbs had adapted as external gills. He called them a missing link, relating amphibians and fish.

These bright suggestions were lost on the few who were interested. Society was not ready to accept any such interpretation involving any kind of evolutionary lineage. Hunter was also largely ignored by his contemporaries for another display of a related lineage – drawings of primate skulls showing the line from monkey to ape to human. 'Our first parents, Adam and Eve, were indisputably black' he wrote, changing over 'thousands of centuries' rather than during the forty-day flood. The Royal Society editor asked him to change the time interval to 'thousands of years' and when Hunter refused the manuscript was rejected. It was eventually published, posthumously, a few months after Darwin's *Origin*, by the Royal College of Surgeons. (Later, the college would urge a medical benefactor, George Buckston Browne, to purchase Down House and they maintained the house and garden there until 1996.)

The Linnean system in Banks's garden had worked well initially, until the project hit an unexpected snag. Solander found the local temptations of Soho society too great and he began to spend more and more time with the young women of the capital.

The growing pile of unclassified herbarium sheets, used to preserve the plants and store them in catalogued cabinets according to Linnaean categories, was left unattended. The problems of classification were then abandoned, to wait almost half a century for another less intrepid explorer, the young Joseph Hooker, to focus on the taxonomy without the local distractions. He added his collections from Australasia and India to those at Kew Gardens, where his father was Director. They worked with the staff at Kew to devise new classifications, while questions of their origins were soon to find their way into the conversations at Down House.

Joseph Hooker's first meeting with Charles Darwin was in 1839, in Trafalgar Square:

> I was walking with an officer who had been his shipmate for a short time in the *Beagle* seven years before . . . I was introduced; the interview was of course brief, and the memory of him that I carried away and still retain was that of a rather tall and rather broad-shouldered man, with a slight stoop, an agreeable and animated expression when talking, beetle brows, and a hollow but mellow voice; and that his greeting of his old acquaintance was sailor-like – that is, delightfully frank and cordial.

Hooker became a frequent visitor to Down House. Some years later Hooker wrote out some of his memories of the times. 'A more hospitable and attractive home under every point of view could not be imagined,' he was to recall. 'There were long walks, romps with the children on hands and knees, music that haunts me still, Darwin's own hearty manner, hollow laugh, and thorough enjoyment of home life with friends.' He took his work with him and stayed for weeks on end:

> . . . enjoying his society as opportunity offered. It was an established rule that he every day pumped me, as he called it, for half an hour or so after breakfast in his study, when he first brought out a heap of slips with questions botanical,

> geographical, &c, for me to answer, and conclude by telling me of the progress he had made in his own work, asking my opinion on various points. I heard his mellow ringing voice calling my name under my window – this was to join him in his daily forenoon walk round the Sandwalk. On joining him I found a grey shooting-coat in summer, and thick cape over his shoulders in winter, and a stout staff in his hand; away we trudged through the garden, where there was always some experiment to visit, and on to the sand-walk, round which a fixed number of turns were taken, during which our conversation would usually run to foreign lands and seas, old friends, old books, and things far off to both mind and eye.

Darwin and Hooker argued long and hard, wrote endless streams of letters to one another about the differences of setting out an evolutionary scheme and a workable classification. When Hooker was sorting through the hundreds of specimens of flowers he had collected in New Zealand he was 'lost to know how to draw the line between there being only one species or 28'. He wanted to limit the number to something manageable and he didn't want the names to keep changing when some bright young botanist came along with some fancy new idea.

Darwin, on the other hand, wanted to show the shared but flexible features between close relatives and, more importantly, how the species had evolved through time from a common ancestor. Yet both men realized the fault at the centre of Linnean's taxonomy: that it was wholly subjective in deciding which characters should be given more weight than others. Hooker searched for a fixed definition based on structural or reproductive characteristics, while Darwin looked at the same features for signs of potential variability.

With all the important new gardens and institutions now following Linnean classification, it was incongruous that the greatest breakthrough in biological thinking should happen elsewhere. What is more, it took place in just an ordinary country house with no collections or equipment, except for a microscope and a postal service. Once again, Darwin's granddaughter Gwen Raverat captures the atmosphere of the place in all its ordinariness:

> The magic began from the moment . . . the door was opened [and] we smelt again the unmistakable cool, empty, country smell of the house, and we rushed over the big, under-furnished rooms in an ecstasy of joy. They reflected the barer way of life of the early nineteenth century, rather than the crowded, fussy mid-Victorian period. The furnishing was ugly in a way, but it was dignified and plain.

But there was another way of looking at what Down House had to offer. Fifty years before Raverat captured these memories of the house, just after the Darwins had moved in, the atmosphere was very different: cold not warm, forbidding not welcoming. They were antagonisms that stretched to other things that Darwin was thinking about in the 1840s, other forces in life. For the moment, he could not make up his mind.

3

A Slow Start at Down

Darkness, depression and doubt

The first three months in their new home had not been good for Emma and Charles and as the last leaves fell that autumn they breathed sighs of exhaustion. The house was cold and still uncomfortable, piles and boxes of their belongings were still unsorted and no one knew where to find anything. They all kept opening unfamiliar doors and bumping into servants. Meanwhile the untouched garden fell into its routine winter quiescence, the harsh weather oblivious to the human confusion.

From the start of the move Emma had feared that she was going to experience the worst of the adventure; but she hadn't even dreamt that it would be this bad. The year 1842 had started very well and her third pregnancy in the fourth year of marriage had made her feel positive and glowing. But then, in November, the baby girl died and her feelings were hard to share. The parents chose to be positive: 'Our sorrow is nothing to what it would have been if she had lived longer and suffered more.' Meanwhile, to add to her troubles, Emma's father, Josiah Wedgwood, had become seriously ill with 'terrible shaking and restlessness' and he was expected to die at any time.

The bad experiences continued with another unexpected event. Early one evening, Bessy, the new nursemaid, had taken the two surviving children out to play along the Sandwalk. For a moment her attention had wandered and when she looked up William and Annie were gone. She searched for more than an hour, and by then distraught she went to tell Emma. After another couple of hours searching, Darwin and the butler Parslow found the two children safe and well, still playing together in the Big Wood beyond Great Pucklands Meadow.

The previous year, Charles had first set down his thoughts on natural selection. Now, in 1843, he began to consider further the question of the adaptation of species: how over time species transform. Being creative about how the process might work, his thoughts moved in two directions. One sought to find how variation came about, why different species evolve and whether the environment had an impact on biological forms. On the other hand, he pondered how species were descended and whether the variation is transmitted from one generation to another. He was particularly interested in finding ways to test these ideas. He hoped that breeding under domestication would produce some answers and this was one of the main reasons why he was so pleased to have moved to Down House, where he could try out as many experiments as he wanted.

Inevitably the slow pace of country life led to more thoughts about what he had sketched in his 35-page proposal. While the conversations at his brother's house with some of the major thinkers of early Victorian England still rang in his ears, he thought about population growth and the way that changing environments held numbers in check, about hierarchies of classification and genealogy, slow or catastrophic change, and even whether transmutation happened at all.

They were cold and gloomy evenings that Emma and Charles spent that winter. In the bare rooms at Down House Charles worried about his work. He was afraid that Owen may have been right after all: perhaps all species have fixed designs. Perhaps Lyell's obsession with gradual and uniform change could be causing them both to miss something important. But in the more confident light of day Darwin was convinced that species were not designed according to some fixed plan but, rather, they changed through long stretches of geological time according to whether the environment stayed the same or not. His questions were becoming clearer as well. Was the slow and gradual change something that happened after rapid catastrophic events? What intervals of time are involved in evolutionary change? How was information passed from one generation to another? Why was sexual reproduction so important? Did teams of organisms compete, or was it individuals?

Although Emma and Charles had escaped to the country the influence of their city friends was maintained by their frequent contact with Charles's brother Erasmus in Great Marlborough Street. He was still part of the London society that included Harriet Martineau, George Eliot and Herbert Spencer. It was Spencer the popular philosopher-economist-sociologist who much later was to coin the phrase 'survival of the fittest' and who became so popular with American capitalists.

Darwin could see that there was a connection between the new economics of industrialized countries like the United States and new species competing with their ancestors in a changed environment. In both cases population growth and the means of subsistence were important controlling influences. Their contemporary Thomas Carlyle, the moral historian from Scotland, called this influence a 'dismal science' based on the premise of unending conflict, struggle and change.

These ideas did not leave Darwin once he had settled at Down. Even the apprentice gardener Henry Lettington noticed that the governor was locked in thought. It was he who said:

> Oh. My poor master has been very sadly. I often wish he had something to do. He moons about in the garden, and I have seen him stand doing nothing before a flower for ten minutes at a time. If he only had something to do I really believe he would be better.

Yet Darwin could not be in total retreat; there, he combined contemplation with the great ideas of the age. Yet, what were thought of in the 1840s and 1850s as major historical events, such as Napoleon III's threats which so disturbed the politicians and the gentry, now have less import. Equally, what were considered to be minor disturbances have greater significance in the long term – like the burrowing of earthworms which largely went unseen. The quiet gentleman of Down House had worked out that he had 53,767 worms to the acre and there was only one Napoleon. Uneventful and unorganized, the wildlife of the countryside continued without heed to seemingly great external events. The transmutation of species, Darwin decided, had no end and no purpose; it continued regardless of the rise and fall of empires.

Darwin's garden became a quiet place for isolated contemplation where all the implications of his concerns on transmutation accumulated. Just as he was arguing that evolution had no end, other thoughts about the meaning of life would go through equally radical transformations. Of course, the world outside wouldn't wait to let Darwin live at his own pace. Events and the challenges from beyond the garden caused him to stop and reappraise his conclusions. He remained a busy man, not the lazy laid back figure some biographers make out. There was the book on volcanic islands to finish, work that he'd started soon after he left the *Beagle* seven years earlier. At Down he had a good study in which to work and he finished his *Journal* of the voyage of the *Beagle*. His other book, *Geological Observations on the Volcanic Islands visited during the Voyage of HMS Beagle*, was published in 1844 with good accounts of coral reefs and island biogeography. But the book made little impact on the many geologists working on volcanoes and made its author regret having spent the eighteen months writing it.

The Lyells and corals

At Down, Darwin began planning a series of projects involving breeding pigeons and primulas, observing the natural process and hoping to uncover some law concerning the transmission of characters (see Chapters 6 and 11 for details of these experiments). He also searched locally for telling interactions between very different species in the countryside. He started with an open mind about how long it would take before answers appeared and even whether the answers would be as clear and precise as the laws of physics and chemistry. The garden began to fill with contraptions to measure the rate of growth of animals and plants, netting and muslin sheets to keep off pollinating insects, cloches to control plant growth and all the accoutrements to keep pigeons and even rabbits.

As was the fashion in middle-class Victorian England, the Darwins entertained many weekend visitors. Darwin was quick-witted and jovial and enjoyed social occasions when his health allowed. Among the couple's closest friends were Sir Charles and Lady Lyell. Sir Charles was one of the landed Scottish gentry and had developed a career as a London barrister. His *Principles of Geology* that had so influenced Darwin was littered with legal rhetoric. The influence of the Scottish Church, with its preaching to explain highland landscape by the catastrophe of the flood had stimulated his rhetorical instincts: 'Never was there a dogma more calculated to foster indolence, and to blunt the keen edge of curiosity, than this assumption of the discordance between the former and existing causes of change.' Instead we need to 'see the ancient spirit of speculation revived, and a desire manifest to cut, rather than patiently untie, the Gordian knot.'

Lyell personified the power of scientific progress in the middle of the nineteenth century and expressed it through his idea that the earth changed slowly and uniformly. The huge coral reefs that Darwin had seen off the coasts of South America were his first proof of Lyell's ideas. They were a good example of whole ecosystems transforming by no more than an inch a century and of how so many different species depend on one another. The raised beaches he had seen along the shoreline in Patagonia might be proof of the same slow change in sea level and could explain similar beaches around the shore of several Scottish lochs. (Both Darwin and Lyell were attracted to this concept of steady, firm geological progress. For Darwin, Lyell offered a model for biological change.)

The country weekend conversations put their excitement about ideas of biological and geological change into a much wider social and political context. Progress was not expected to be towards a fixed goal, as in religions, or even in Jeremy Bentham's utilitarian world of the 'greatest good for the greatest number'. Lyell was one of the few thinkers who had any time for unpredictable processes let alone non-directional ones: to most others they were sinful, immoral or unscientific.

Darwin's growing friendship with Lyell would be tested by the rigours of harsh scientific scrutiny for decades to come. Even at the start they found themselves with different explanations of how coral reefs formed. Lyell had written about their formation in his *Principles of Geology*, arguing that they grew slowly around the edge of volcanic islands that were slowly sinking under the sea. The coral could only live a few feet below the warm and light surface of the water and they grew into reefs hundreds of miles long and as many feet deep. They were just another part of the gradual change in the earth's structure, volcanoes and coral slowly building new land masses out of the sea. However, further evidence from surveying voyages such as that of the *Beagle* offered another explanation. Depth soundings showed that the reefs rose up many hundred feet straight from the sea floor, with no volcanoes in sight.

Lyell faced up to the failure of his theory in a letter to the astronomer Sir John Herschel:

> I am very full of Darwin's new theory of Coral Islands, and have urged Whewell to make him read it at our next meeting. I must give up my volcanic crater theory for ever, though it cost me a pang at first, for it accounted for so much, the annular form, the central lagoon, the sudden rising of an isolated mountain in a deep sea.

The 1844 essay

When his weekend guests had left, Charles would return to his work in solitude. It was in the old study at Down House that most of his planning and writing took place. As in most homes, the fireplace was the focus, with shelves for books and papers on the right-hand side and a marble-topped table with a basin and water jug on the other. In the middle of the room a large table was piled with books and specimens, microscopes and notebooks. Here he sat writing and often when he lost concentration he jumped up and walked to another table in the hall on which stood a jar of snuff. He took enough to set himself to work again.

It was an exciting challenge to sit there and reconsider the 35-page essay on transmutation that he had scribbled out two years earlier. Indeed, he spent several months, on and off, writing another version and he didn't finish it until the summer of 1844. The central theme was the manner in which organisms changed in different environments and how this was a central means of change from one species to another. If the environment changed, those that happened to be adapted to the new conditions would survive at the expense of the others. He was also thinking about the limit on the numbers of individuals in an expanding population and whether the marginalized individuals would survive beside the others.

This 231-page essay of 1844 emphasized the evolutionary tree of common descent and the importance of geographical migration. The chapter headings established the pattern of topics well known in the later versions of *The Origin of Species* and in this first version they were split into two parts: 'On the variation of Organic Beings under Domestication and in the Natural State' and 'On the evidence favourable and opposed to the view that species are naturally formed races descended from common Stocks'.

Through the months of writing in those first few years at Down Charles became more anxious about the impact of his conclusions, the reaction of his friends and the whole scientific community. He was only thirty-five years old and not quite sure enough of his ideas and the new ways of science to be too confident. But he was aware that he had enough time in life to expect some further breakthroughs. He knew he could get more evidence and it was much more in his character to wait for the results to emerge and to work hard to get them, rather than to publicize an untested theory without enough evidence. His main message in the sketch was the right argument to defend his proposition scientifically: 'I cannot possibly believe that a false theory would explain so many classes of facts, as I certainly think it does explain. On these grounds I drop my anchor, and believe that the difficulties will soon disappear.'

In July 1844 he sent his scribbled pages to be copied 'in a clerk's hand' and it is this manuscript that survives. In case he died before the work was published he set out plans for the work to be respectfully edited and published in a formally written note to Emma:

> I have just finished my sketch of my species theory. If, as I believe that my theory is true & if it be accepted even by one competent judge, it will be a considerable step in science. I therefore write this, in case of my sudden death, as my most solemn & last request . . .

In the autumn of 1844, just as he had finished the extended draft, an anonymously written book was published in Edinburgh that gave Darwin and thousands of others an unexpected fright. The book was called *Vestiges of the Natural History of Creation* and argued that evolution progressed up a ladder of complexity, but without the influence of God. New habits led to altered structures rather than the other way round. The secret author of *Vestiges* challenged the Creationist view that when the first birds suddenly found themselves with a wing they soon found something to do with it. The *Vestiges* then went on to suggest that the major differences between one large group, such as birds, and another large group, such as mammals, happened at the embryo stage – all in one sudden jump.

The mystery author was praised by the critics for being a very good writer and the book was also expertly marketed. It quickly became a bestseller and ten editions were published. The book was extremely critical of the religious establishment and perpetrated numerous errors and misunderstandings of nature. This angered the scientific establishment, most of whom were churchmen. But the book's popularity only made the matter worse.

The principal objection to *Vestiges* was its suggestion that progress took place without God's design, implying that nature evolved of its own accord. This was all too much for one reviewer, the Cambridge geologist Reverend Adam Sedgwick. Here was some cowardly bungler attacking all his life's work and belief: 'From the bottom of my soul I loathe and detest *Vestiges*.' In the spirit of the time he dismissed the work as trivial, saying that anything so stupid must have been written by a woman.

It was quite common for books to be published without giving the author's name, and this stimulated added interest in trying to guess who the author was. Among the candidates were the novelist William Makepeace Thackeray, Harriet Martineau, Charles Babbage, Charles Lyell, Charles Darwin and even Prince Albert. The owner of the influential *Westminster Review*, John Chapman, was in a better position to know than most and in 1848 he assured Richard Owen that 'there is now pretty strong evidence to fix the paternity on Chambers'. Robert Chambers was the founder of the Scottish publishing company that specialized in encyclopedias. But Chambers kept silent and covered his tracks, having a pew in two churches so that people seeing him absent from one would assume he was at the other.

Darwin felt sick at the very mention of 'Mr Vestige's book'. The reaction to the book had made him fear the clamour that would respond to his own ideas, and he decided to put publication of his own book on hold. For the moment, at least, he had to remain on the defensive: 'The geology is bad and his zoology far worse' he wrote to Joseph Hooker at Kew. But his friend gave a different response: 'I have been delighted with *Vestiges* from the multiplicity of facts he brings together [even though] he has lots of errors.' The episode made it clearer than ever to Darwin that he had a lot of work to do gathering support from scientific specialists, making his own observations and presenting the argument so that both specialists and the public could understand.

For Darwin's thirty-ninth birthday in February 1848, Emma gave a weekend party at Down House. It was a very male affair with much talk about science, and the ladies joined them on nature walks around the village and across to the Lubbocks' estate. As well as Sir Charles and Lady Lyell, Emma invited Richard Owen, the young geologists Edward Forbes and Andrew Ramsay, and another bachelor, a botanist called Robert Brown. Brown, the greatest botanist of the day, who worked at the British Museum, is best known for discovering what is now called Brownian movement, a self-organized reaction of small particles under tension. The party was the perfect opportunity for Charles to test his developing thoughts about natural selection on other naturalists.

On this occasion the after-dinner conversations between the scientists touched on the connection between transmutation of a particular species and the stability of the rest of nature. Were the catastrophes of the Peruvian earthquake signs that Lyell needed to rethink his theory? His slow and steady rhythms might undergo a change of tempo from time to time. The group argued about whether everything in nature was planned and whether everything was useful. Most agreed that there need not be an aim, nor an integrated stable state to the whole system of nature. They expected an interplay of separate purposes, some competitive, some cooperative and others commensal. Yet Darwin did not get the full support of his colleagues.

According to her journal, Emma shared her thoughts about the party with her daughter Bessy:

> Mr Lyell is enough to flatten a party as he never speaks above his breath, so that everybody keeps lowering their tone to his. Mr Brown whom Humbolt called 'the glory of Great Britain' looks so shy, as if he longed to shrink into himself and disappear entirely; however, notwithstanding those two dead weights, viz. the greatest botanist and the greatest geologist in Europe, we did very well and had no pauses. Mrs Henslow had a good, loud, sharp voice which was a great comfort and Mrs Lyell had a very constant supply of talk.

Emma shared her character assessments with Charles and they agreed that the group had been very stiff. When things go slowly they often need a kick start to move forward.

As though in response to these observations, ten days after the party the French king abdicated and sought exile in England. Talk of catastrophic change was more topical in politics than it was in science. At another party Lyell and Owen were with Sir Robert Peel, the Tory leader, and heard that a new republic was to be declared in France the next day and that there were 30,000 communists 'who were for property in common and no marriage and who are much to be feared'. The English authorities made precautions, the Queen left London; 85,000 special constables were recruited and 7,000 troops mobilized. Some alarmists expected 150,000 Chartists at Kennington Common. At least Lyell could tease Owen that it was all brought on by the French geologists' faith in catastrophe.

The Darwins and religion

The argument for slow and gradual transformation as an explanation for evolutionary change was very sensitive in nineteenth-century Europe. It touched on religion as well as politics and science. Most people had learnt to accept the work of a Creator to explain the beginning of life on earth. They were doubtful of this more complicated biologically driven explanation of the origin and diversification of life. If it were to be gradual, vast amounts of time were needed and it was not clear whether life had been around that long. The alternative drive from catastrophic events, that some suggested might have stimulated evolution, was shrouded in even more uncertainty. Would volcanic catastrophes, such as the eruptions Darwin and FitzRoy saw in Chile in 1835, help speed things up or slow things down?

Very different feelings about religious faith were to become an important barrier between Charles and Emma, especially after the deaths of their young children. Emma retained her belief in God as a source of love and goodness and she remained an active member of the local congregation, running the Sunday School for the village children and regularly visiting the 'Good Poor' of Downe, as the middle classes did then.

During these years Darwin was working at Down House trying to sort out what to do and how to do it. The *Beagle* journal was published. The fuss over the arguments about evolution in *The Vestiges* had forced him away from his 1844 manuscript for the time being. In 1846 Darwin wrote to Hooker with plans to begin work on barnacles: 'I daresay it will take me five years, & then when published, I daresay I shall stand infinitely low in the opinion of all sound naturalists.' What was needed was more scientific credibility and that was why he focussed his mind on barnacle structures and life cycles. He saw it as an investment, a few years of work that would give him the status he needed for the people that mattered to listen to him seriously.

Charles's faith was further shattered when his father died in November 1848. It had a terrible effect on Charles. He couldn't attend the funeral and spent days, weeks, shuffling around the house and garden. Dr Robert Darwin's estate left Ras and Charles rich men but this was no compensation for the mysterious love and unspoken bonds between them. Darwin expressed some of his reaction by anger with God:

> I had a slow realization that 'disbelief' had crept over me at a very low rate but was at last complete. The rate was so slow that I felt no distress, and have never since doubted even for a single second that my conclusion was correct.

They were also difficult times with his children's health and at one time in 1849 all three girls had scarlet fever. The eldest, Annie, had been born in Gower Street a year before the move to Down House, and she loved the garden and its country life. In November 1850 she wrote to a friend:

> We have a new pony. It is rather a little one. I think your donkey sounds a very nice one. I should like to see your little white guinea-fowl. Are all the little turkeys sold? On Sunday

> it was raining dreadfully, and the pit in the sand walk was full of water. Is your swing taken down? Ours has been taken down a long while.

Annie didn't get better and the symptoms reminded her father of 'an exaggerated one of my . . . illnesses. She inherits, I fear with grief, my wretched digestion.' Annie died in the spring of 1851 at the same Malvern clinic where her father had sought a cure for his own gastric illness. The tragedy added to Emma and Charles's woes, taking Emma closer to the Church and Charles to his theory.

For Charles Darwin life would be explained by science not doctrine. Evolution would be understood by experiment, leading to a better understanding of medicine and agriculture. At the time, some hoped its answers would also provide a solution to Malthus's challenge of sustaining the population. Understanding the process of evolution would also be the way to explain the meaning of life.

Meantime, there were several things missing from the two draft essays. One was a link between the two major interacting systems, the organic living systems and the outer inorganic environment. Another detail missing was a link between one generation and the next, some kind of transmission factor presumably through sexual reproduction. How does the inside of a cell, or a whole organism for that matter, communicate with the environment? And how are the changes noted and passed on for the next generation to use?

It was going to take a lot of courage to face up to these challenges in public. To be taken seriously he had to develop plenty of knowledge and understanding, to grow strong like the oaks and beeches in the copse.

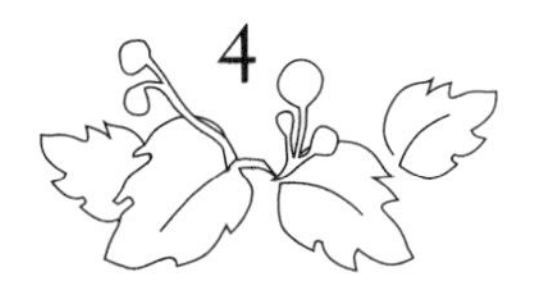

4

The Tree of Life

Growing in confidence

Darwin's painful family history haunted him, and his own guilt about staying alive encouraged more worry about his own fragile health. In the same way it meant that committing himself to anything likely to be controversial with the scientific establishment became very slow work. The uncertainty got worse when he was confronted with the major decisions of his life like getting married or publishing his theory of transmutation. It especially tormented him when he had confronted the failing health of his dearest daughter Annie. He approached all these dilemmas in the same way, by drawing up a balance sheet.

Darwin needed to find an order to his ideas. Instead of the two-by-two march on to the Biblical Ark and the two-dimensional order of Bonnet's ladder, he favoured an older analogy to show another way of thinking about the progression of species: a branching tree, the Tree of Life. The branches grew in straight lines, diversified and separated at different rates. A lineage branched off from the ancestors as a new species with an independent fate. He liked to consider complex process as a whole, but the image of the branching tree offered a useful model to show the changes through time as well as structure.

Darwin had made a little sketch in his private 1837 'Transmutation Notebook' of how he saw evolutionary progress, one of the first ideas he noted down after he returned from South America. It showed four main branches, which he called genera, from the same origin. He drew each with a terminal fan made up of between three and seven fingers, which signified species. Originally the branching tree had been a popular icon used by ancient cultures in Egypt, Israel and Greece; the drawing showed how things split in two or more directions and accommodated differences, and could help make decisions when faced with a dilemma like whether to get married or not. It also allowed the possibility that lineages might diverge. Darwin's sketch was to become the crux of his presentation of the theory of evolution and he used the image to illustrate his theories in a presentation to the Linnean Society, twenty-one years later. It was an optimistic celebration of life's diversity.

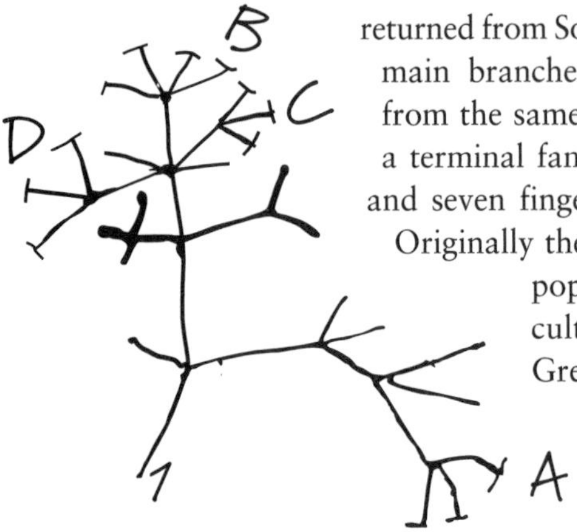

From this sketch of a branching tree, Darwin developed the single diagram he later used in the 1859 *Origin of Species,* which showed how different varieties diversified from the same species. The example showed eight species that evolved through at least 14,000 generations from one origin. Each species produced a fan of variants, another branch in the tree. It was always the most extreme, the most marginal that led on to the new branch, while the mainstream species either just went on without diverging at all or became extinct.

Darwin had observed that the more numerous a species, the more variable its range of form. With many individuals sharing the same features there was more opportunity for natural selection to diverge and lead to more species, thus increasing the complexity of the web of life, the interactions between species and environments, the different developments within different generations. Unlike his grandfather, however, Charles did not see the process leading along the branches of the tree to any

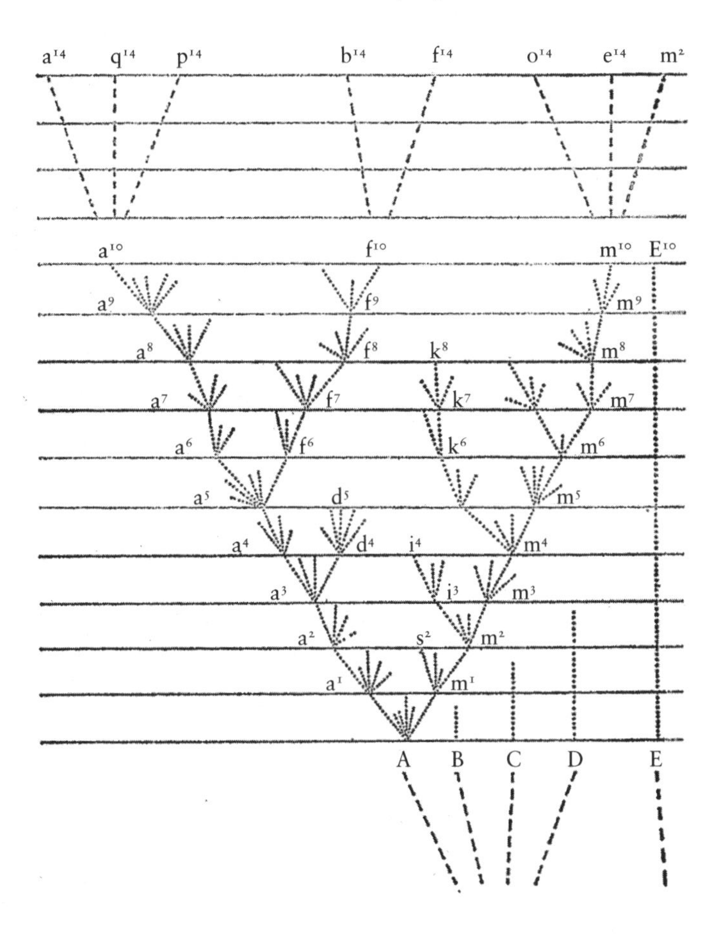

particular target by one particular mechanism. Controversially, there was not one supreme goal – man was not the pinnacle of progression, as had been supposed by Bonnet and in the Chain of Being – and there was not necessarily one mechanism – a Creator – for evolutionary change.

But there was a lot of related talk at the time, of things having opposites and becoming separate, most famously from Hegel's *Science of Logic* that had just been translated into English. Darwin's comparison to a mother giving birth to two different children was less alarming to most Victorians. This action, of one species splitting into two, also offered other ways of thinking about species, conflict, advancement and what happened when there was no progress. Such thoughts enabled vigorous conflict between two systems in which neither side really won. The mother species became two daughter species. If nature was to change then it must be continually balancing one option with another, continually elaborating on all the new opportunities.

This way there was no need for a target of any other kind of established model of the kind Cuvier and Owen expected. At one of his lectures in 1837 Owen had asked the audience, which included the eighteen-year-old Queen Victoria as well as Charles Darwin, 'Can the various structures which comparative anatomy now unfolds, be referred to one, or do they manifest different types? This is a question which is now in progress of solution.' Darwin's answer was going to be very different to Owen's; for him transmutation was an open process of development.

Nevertheless, the relentless process did form a hierarchy as it proceeded from one form to another. Each level presupposed the earlier ones, sometimes building something new on top of the other. Biology was not like Newton's physics but was an interactive network and it kept on developing. Darwin enjoyed playing with these thoughts privately and slowly he became confident about the way organisms adapted. Surviving organisms displayed a spectrum of variation for each feature, diverging when environmental change made that necessary. Otherwise the struggle ended in extinction.

This idea of progress involved many fuzzy branches in the Tree of Life which were difficult to trace back to their time of origin, and this led to many conflicts and misunderstandings. One of the first came in the 1840s. In the spirit of neighbourliness, Charles began to teach John Lubbuck, the squire's son, how to use a microscope. Darwin's eldest son, Frances, had also been in thrall to nature and with such a charismatic teacher the teenage boys soon became competent naturalists. Lubbock drew well and Darwin helped him get the really accurate illustrations published, starting off the young man's career as a professional naturalist. Both teacher and pupils were looking at many things for the first time. There was no species name for many of the specimens of water fleas they had collected, maybe only a genus or Family name or even a common name. The fuzzy lines between potential species seemed to merge the range of biological structures in these and other organisms. The three Down House naturalists were left wondering whether they had found the right clues of where evolution had led.

The young boys were often joined at Down by the most frequent and favoured botanical visitor Joseph Hooker, and they developed a strong and affectionate relationship with him. Francis and John knew that Hooker had a secret liking for gooseberries and often caught him helping himself to the summer fruit from the Kitchen Garden. A large part of this productive patch was taken up with hundreds of gooseberry bushes and Darwin was proud to have more than fifty varieties. They hybridized easily and showed variations in growth of the shape of the bushes as well as the fruit. The existence of so many varieties in the 1840s and 1850s showed that evolutionary change within these artificial populations could happen within a few generations. How this happened in the wider natural world would take longer to confirm.

Lessons from barnacles

Through the middle of the nineteenth century it was fashionable for one naturalist to take responsibility for just one branch of the Tree of Life, one Family or Order of animals or plants. Most enthusiasts specialized with their one group in their own region while only a few really ambitious scientists offered a global perspective. As a result the territory of biology was a hazardous one in which the knowledge of species was uneven. The larger mammals and the most nutritious or beautiful flowering plants were among the first to be studied, leaving most other species and genera poorly described. Some groups were reviewed systematically by a small number of enthusiasts, such as the vertebrates which were studied by Cuvier and Owen, jellyfish by Thomas Huxley and barnacles by Darwin, who spent six years at Down House comparing hundreds of species.

Darwin had collected many barnacle specimens himself at the many ports of call made on the *Beagle* expedition and the investigation that followed was done in the study at Down House. There, he also looked at hundreds of other kinds of barnacle specimens collected by others from all over the world, sent by post in little pillboxes. As well as the piles of papers and books on the big table in the middle of the room, there were half-finished sketches and notes of ideas, open envelopes and packages. During this period there were also more unusual pieces of equipment such as a wooden dissecting board, knives, scalpels, pins, even a little hammer, all scattered around the two microscopes. One was for dissecting and looking at surface structures, the heavy brass one for higher magnifications of up to 500 times. For that instrument he needed objects thin enough to enable light to pass through them, mounted on glass slides and stained with bright colours to show the different chemical structures.

With the barnacles Darwin had taken on a most varied group. Many of them had a free-swimming stage to their life cycle that was difficult to recognize. Their volcano-shaped shells became fixed to rocks exposed by the sea or to the bottoms of boats. Some of the sedentary forms had stalks, others burrowed in the substrate with the help of specially secreted chemicals. To name and classify these different animals Darwin had to investigate all these characteristics and more beside, and the work tested his skills of observation, analysis and dissection to the limit.

As he discovered more about the range of anatomical detail in this very diverse group he could see more clearly that there was an underlying body pattern. Some species developed particular parts of their bodies more than other parts and Darwin saw that this was how each species adapted to its own particular environment. Thus he classified the hundreds of living and fossil species into three main branches: acorn barnacles with a conical shell of different numbers of plates; goose barnacles with a muscular stalk allowing the body to hang down into the sea to feed; and a group with no hard parts which was parasitic on animals like crabs. Most of these differences could be explained as divergences from close relatives, and some of their organs had changed their function by differing amounts to meet different and new conditions.

The laborious work stretched even Darwin's patience and in 1849 he wrote to Lyell to share some of his frustrations, concluding with the opinion that an 'Omnipotent God made a world, found it a failure, and broke it up, and then made it again, and again broke it up, is all fiddle faddle.' Then Darwin described his feelings about his own work on the barnacles:

> I work now every day at the Cirripedia for 2 ½ hours, and so get on a little, but very slowly. I sometimes, after being a whole week employed, have described perhaps only two species; however, the other day I got the curious case of a unisexual, instead of hermaphrodite cirripede, in which the female had the common cirripedial character, and in two valves of her shell had two little pockets, in *each* of which she kept a little husband; I do not know of any other case where a female invariably has two husbands.

Sometimes the dilemma of deciding where one species began and another ended was too frustrating:

> What miserable work, again, it is searching for priority of names. I have just finished two species, which possess even seven generic, and twenty-four species names! My chief comfort is, that the work must be sometime done, and I may as well do it, as any one else.

And then, in the same letter to Lyell, he expressed his conflict of interests between the two kinds of hierarchy. On one hand was the descriptive classification, deciding on which species went in which genus and Family. On the other there were the much more holistic interpretations about the barnacles' own evolutionary tree. This speculation led on to the even greater challenge of the relationships between all animals and all plants:

> I declare your decided approval of my plain Barnacle work over theoretic species work, had very great influence in deciding me to go on with the former, and defer my species paper.

One dissection that was particularly baffling was of a specimen of an undescribed species that had been sent from the Philippines. Barnacles usually have male and female organs on the same specimen but here they were separate and very different to one another's or to anything else. It had taken time and skill for Darwin to realize what they were and more days passed to reveal yet more variation from the normal. He wrote to his friend John Henslow:

> the female has the ordinary appearance, whereas the male has no one part of its body like the female & is microscopically minute; but here comes the odd fact, the male or sometimes two males, at the instant they cease being locomotive larvae become parasitic within the sack of the female, & thus fixed & half embedded in the flesh of their wives they pass their whole lives & can never move again.

Darwin knew this work on barnacles would involve lots of hard work but by Christmas 1849 this goal still seemed a long way off and his stomach cramps and vomiting returned. He spent more time in his study, where he absorbed himself with more new barnacle specimens that were being mailed to him from friendly correspondents all over the world. The work had plenty of rewards, for the range of form and the ways in which these creatures adapted to great variations in living conditions was quite astonishing. He found one species that emitted its excrement from its mouth because it didn't have an anus. One tore its food with toothed ridges on its legs. Another had teeth inside its oesophagus and had no thoracic legs at all. A stalked species had its very long penis coiled up like a spring while in another species there were two smaller straight ones.

It was proving to be a hard and strenuous job, with hundreds of species from all over the world, always needing microscopic examination, always demanding a critical and open mind. Yet from his work on barnacles he concluded that a 'community of embryonic structure reveals community of descent'. Development seemed to be playing a part in structuring a body according to some underlying plan: by growth in the embryo rather than by Creationism. In England this was dangerous talk, for it confronted belief in the organization of nature according to God. Owen and the French, with their belief in the fixity of species and catastrophic change, were waiting to pounce so Darwin kept quiet.

The first of four volumes on the barnacle work, published in 1851, was entirely descriptive and left controversy until later. Here there were not only many new species, but many of these in turn fitted into new genera, Families and Orders. More than that, Darwin showed that this very difficult and diverse class was part of the Crustaceae Phylum, not the Mollusca.

But the barnacle work did not go down well with his friend and supporter Joseph Hooker, who was collecting plants in northern India in 1849 and had become tired of all the letters about barnacles coming from Down House. He wrote and told Darwin so, pressing that the species book was more important, but admitting that perhaps it might be best for them both to find out more about the gradations of varieties within a species. Hooker's Indian plants were much more concise than the barnacles, but then the plants' environments were much less varied. It seemed as though evolutionary change happened at very different rates, depending on the changing environment outside the organism and inside its cells.

For Darwin it seemed increasingly important, however, to study whole life cycles of the entire group and the development of each part. There were several benefits of seeing the project through to the end, not the least of which was that it allowed his ideas to mature. The study of barnacles also challenged his powers of observation and showed just how fine the variation in these organisms was. The project stretched the power of his shining brass microscope to the limit, showing cellular detail from the delicately adapted limbs and guts of these highly varied and diverse creatures.

After Darwin's death Thomas Huxley told his son Francis what he thought the value of the years given to the Cirripedes had been:

> In my opinion your sagacious father never did a wiser thing than when he devoted himself to the years of patient toil which the Cirripedia-book cost him. Like the rest of us, he had no proper training in biological science, and it has always struck me as a remarkable instance of his scientific insight, that he saw the necessity of giving himself such training, and of his courage, that he did not shirk the labour of obtaining it.

Lessons from the oaks

Early on Emma and Charles had set aside a plot at the southern end of their land by the cricket field, 256 by 164 feet (80 by 50 metres), which they fenced off from stray grazing animals to plant a copse of native trees. Like the barnacles, the trees in the copse were closely related and closely classified. For centuries recurring generations of oaks, hazels, alders, hornbeams, birches and beeches had been common on the North Downs. The beech in particular had dense foliage spreading horizontally on its branches which caused the leaf canopy to stop most direct sunlight penetrating to the ground. They were trees that had been present in modern and ancient forests for millions of years, and many of the species were classified as major parts of the same Order of flowering plants, the Fagales. The features that caused botanists like Hooker to classify these trees and shrubs together were the similar small flowers, catkins, and the pollen that was carried by the wind. Some good specimen trees of the group surrounded the lawns and fields around the garden as well as making up much of the nearby woodland. This was the range of trees that the newly wed couple planted at the end of their garden as one of their first improvements. Charles was chuffed to bring together both the classificatory and the evolutionary meanings of the Tree of Life in the little copse.

When Emma and Charles Darwin were selecting the trees to be planted, these implications of their project were only partly understood. Without the fossil record and DNA sequence data, Victorian taxonomists relied entirely on structural features, their similarities and differences. At that time, hornbeam and hazel were classified in a different Family to birch and elder, while oak and beech had traditionally been put together. In time other specialists would propose different classifications of these groups because the slow discovery of new data encouraged different interpretations from people with different expertise.

As the new trees grew into young woodland the Darwin children habitually played there. During the Crimean War with Russia the boys dressed up in dark-blue uniforms and marched along the Sandwalk to the tunes of George's tin whistle and Frank's bugle, then pitched camp in the copse. The flames of the camp fire cast strange shapes on the trees close by so it wasn't hard to fear the Tsar's troops advancing with fierce cries. Etty and the other girls called this the 'Dark Side' of the copse and were dared by the younger boys to walk through it at night alone. The answer was to take a friendly teddy bear for protection against ghosts jumping out from hollow trees. There was one of the beech trees that was especially spooky with scars on its broad trunk so that in the shadows it looked like an elephant.

The 'Light Side' of the copse looked out across Great Pucklands Meadow to the woods beyond and it was here that the summerhouse offered quieter family pursuits – reading and knitting were favourites for Emma and the nurse Brodie while they looked after the children. The Light Side was where the hedgerows gave a special kind of unnatural habitat, tended by the gardeners. It was an ecosystem of selected species most of which were specially planted. The hedgerow at Down had dogwood and privet, anemones, primroses, cowslips, ivy and holly. It also attracted other herbaceous species, seasonal wild flowers and weeds such as hedge parsley splitting into many closely related varieties and species. The many different forms of these wayside herbs made some botanists of the day wonder whether they had specially adapted to the human-made locations.

Etty remembered a day just after a local lad had been given

a job to pull out the poisonous dog's mercury that had invaded the copse:

> As my father and mother reached the Sandwalk they found bare earth, a great heap of wild ivy torn up by its roots and the abhorred Dog's mercury flourishing alone. My father could not help laughing at her dismay . . . and he used to say it was the only time she was ever cross with him.

Looking for clues

Polly the terrier, Tommy the cob and the many pigeons whose names are not recorded all made it difficult to run an ordered house and garden, and Emma was always tidying up after Charles. Like his plants, however, these creatures were being bred for special reasons, some loving, others more functional and most experimental. Their variations raised the possibility that it was chance that led to the branching of the Tree of Life, with individual lineages, not general laws, controlling nature. Darwin's strongest hint at this was to come in *The Origin of Species*: 'I believe in no fixed law of development. The variability of each species is quite independent of that of all others.' The very existence of such diversity in Darwin's garden, the small differences in so many similar species, was also due to the changing environment. To keep the system in equilibrium, nature seemed to Darwin to have a combination of self-organization, hereditary programming and natural selection which drove the beautifully integrated diversity of biology. But how could anyone observe all of this? There were plenty of clues in the garden showing how it was done and only a few were to escape Darwin's eagle eyes.

He watched as bees pollinated in adverse conditions because they wanted nectar and it was not always available. Any pollen picked up on the way was incidental. Lime, dogwood and privet, anemones, primroses, cowslips, ivy and holly all grew together in the hedgerow where they adapted in size and colour to make new varieties and hybrids. More rarely, the breeding of some strange animal broke through the species barrier in new environments to favour the survival of the unconventional, like the missel thrush surviving over the song thrush or the Asiatic cockroach over the native Russian species.

Following his experience on the *Beagle* voyage, Darwin was always on the lookout for such obscure signs of origin and change in plant and animal species. Evidence for the migration of familiar species over short and long distances was another thing he watched out for and the chalk escarpment at Downe had many species that he had encountered in different forms on his voyage. It was as though these related species were temperate outreaches of groups with headquarters elsewhere.

With his knowledge of the Kent flora and fauna, and with his work as Secretary at the Geological Society in London, Darwin had a good idea of how these plants and animals had come about. Through the geological and climatic events that had shaped the Wealden landscape down to the south coast, the top of the North Downs escarpment where he lived had seen big changes. Victorian geologists believed that most of these processes were very slow, assumptions influenced by the ideas of Charles Lyell. It was impossible however to be sure of these assumptions. Their conclusions were based on unreliable measurements of the rate of sedimentation that put all estimates of geological time into the realms of speculation. They also assumed that the processes were gradual and proceeded at the same rate. For the first edition of *The Origin* Darwin calculated that 'the sea would eat into cliffs 500 feet [152 metres] in height at the rate of one inch in a century . . . At this rate the denudation of the Weald must have required 306,662,400 years; or say three hundred million years.'

These judgements were estimates of long and lazy time, more in line with the quietness of life at Down House in the 1850s than with any scientific evidence. William Thomson, later to be ennobled as the first Baron Kelvin, was already using physical measurements from heat pumps to work out the second law of thermodynamics. He went on to use the theory to calculate the age of the Earth and came up with a much shorter timescale that brought the hazy lifestyle at Down House up against another traumatic reality. The aimless Tree of Life was more chaotic than

Darwin and others had thought by just gazing at the two-dimensional figure in his notebook. Meanwhile, his thoughts along the garden path were going on to make another creative leap.

Rather than be impressed by Kelvin's numbers, Darwin was at his most creative as an observer. This passion was inherited by his niece, Gwen Raverat, who caught the atmosphere of the nineteenth-century garden in *Period Piece*:

> . . . the path in front of the veranda was made of large round water-worn pebbles, from some sea beach. They were not loose, but stuck down tight in moss and sand, and were black and shiny, as if they had been polished. I adored those pebbles. I mean literally, *adored*; worshipped. This passion made me feel quite sick sometimes. And it was adoration that I felt for the foxgloves at Down, and for the stiff red clay of the Sand-walk clay-pit; and for the beautiful white paint on the nursery floor. This kind of feeling hits you in the stomach, and in the ends of your fingers, and it is probably the most important thing in life. Long after I have forgotten all my human loves, I shall still remember the smell of a gooseberry leaf, or the feel of the wet grass on my bare feet; or the pebbles in the path. In the long run it is this feeling that makes life worth living, this which is the driving force behind the artist's need to create.

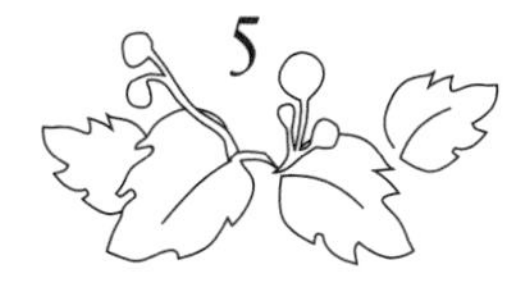

5

Entanglements

Suddenly, Joseph the coachman pulled hard on the reins and the wheels screeched to a halt. The one passenger had shouted out, scaring the driver and causing him to pull into the side quickly. Joseph jumped down and looked up to his disturbed charge to find him laughing, with a wide grin overshadowing his sober beard. Darwin's years of doubt were over and there was no need for intellectual caution. He explained in his autobiography how this experience gave him the confidence to go ahead and publish his ideas on evolution by natural selection and now with the added argument of divergence. For the first time, that May evening in 1855 he was able to answer all the critics he knew would be waiting. 'I know it!' he shouted. 'It's going to work.' He had made the breakthrough he'd been hoping for.

Darwin's new confidence had made him more relaxed and enabled him to experience a revelation from something very ordinary, an unexpected detail that he had observed every day during his routine walk in the garden but had never thought to consider. Hedge parsley was one of the most common flowering plants growing by the fence separating Great Pucklands Meadow and the Sandwalk. The weed liked to grow in the freshly disturbed and open soil of hedgerows and beside footpaths, places that needed human intervention to keep them from being overgrown. Indeed, several species and varieties of hedge parsley were likely to have evolved specially to adapt to this recently available niche. The many closely related species had similar origins in slightly different places: some lying flat on the ground, others crawling over exposed rocks and bare ground. Their differences showed conflict, which was why there were so many closely related species and why they were to give Darwin such an important clue about divergence and evolution.

Darwin's eureka moment came unexpectedly and without fuss from just such observations and their careful consideration. On hundreds of occasions he had passed by the hedge parsley in the garden hedges and growing by the roads of the Downe countryside. This time he was returning from London where he had been attending a meeting at Burlington House and his horse-drawn carriage picked him up as usual from the station at Orpington. The narrow lane was deserted and offered the few travellers that were interested a wonderful chance to look closely at the varied hedgerow insects and wild plants that Darwin loved so much.

It was one of the happiest moments in his life and he felt immersed in a warm and coloured beauty. The sunlight on the brown road and the hedgerow led out to the quiet green landscape – the great complex vista moulded into his thoughts about evolution. All these unlikely things and the dreadful people back in the city, everything was in a delicate balance. He loved the countryside and he hated the city. He disliked all those people and in particular all those ambitious scientists trying to impress him. Now that he'd got some of his final doubts resolved about his latest results he felt relaxed and very pleased with himself. He needed nothing from any of them. His ideas could stand or fall for he had made up his mind and he would never go to another meeting again.

The breakthrough was a simple idea that had eluded him and others, an idea that clarified an earlier unanswerable question. He had spotted a patch of hedge parsley with an unfamiliar flower colour that might be a different species, different to those he saw each day beside the Sandwalk. How could a group like this appear in a different environment? Was this how a new species was created, by moving away into new territory? Thomas

Malthus had suggested more than fifty years before that less healthy individuals would die if they stayed in the same place. That was for humans, in reaction to the serious squalor that faced so many of the workless citizens in the overcrowded cities of early nineteenth-century England. Was the possible move of hedge parsley from the Sandwalk to this green country lane one way in which nature dealt with the same problem? Was it another aspect of the same process of the origin of new species?

Complexity from the Enlightenment

This notion of the interconnectedness of different things, varying from objects to feelings, came from the inquisitiveness of the eighteenth century. Here, for instance, it involved flower colour and migration and the hidden causes of both. At the heart of the enlightened attitudes to nature lay a nest of paradoxes. The desire to discover unspoilt nature was offset by the need of others to improve nature with science and agriculture. Nature became the new religion as well as a laboratory. So William Blake had compared the dark satanic mills in the industrial north of England with the Jerusalem of new opportunity and pleasures opening up in his home town of London. He had celebrated this new society and the ways of life starting there in poetry:

> The fields from Islington to Marylebone,
> To Primrose Hill and Saint John's Wood,
> Were builded over with pillars of gold;
> And there Jerusalem's pillars stood.

At the same time, just south of London, Malthus was warning about the likely implications of these new trends:

> Population must always be kept down to the level of the means of subsistence; but no writer . . . has inquired particularly into the means by which this level is effected: and it is a view of these means, which forms, to his mind, the strongest obstacle in the way to any great future improvement of society.

Utilitarians such as John Stuart Mill and Jeremy Bentham may have wanted more people to work and fight wars and pay taxes, but rising population was an obstacle to affluence.

It hadn't dawned on many of Malthus's readers, fixed as they were by strong social roots, that migration far away from the place of their origins was an option for survival. It was this revelation that flashed across Darwin's mind as he observed the species of hedge parsley. He saw instantly that what was true for the adaptation of new plants was also true for humans in London's East End. Given plenty of time the newly adapted groups would move and thrive. The full consequences of these apparently straightforward influences took time to penetrate the minds of most people. Was it to mean that those left behind were to become extinct? They were, after all, very complex processes happening inside and outside the individual organisms and it was difficult to understand how all the processes from many disciplines came together and interacted. Even our societies, leaders and regional cultures didn't accept that humans were just another species, that our species was no more supreme than any other. We were not in control of our lives or the environment: God had not given us some kind of special licence. The complex system of nature controlled the diversification of species and their evolution and Darwin was pleased with his new hunch about it that evening on his way home to Downe.

He could see that diversification led to the branching of the Tree of Life. The separation of one group from another was a force from within the group once plenitude had been reached, a state of complete fullness within the whole system. As one group of organisms moved away, so it was for that group to first transform as a new variety and then a new species. Darwin saw at that moment how the new hedge parsley by the roadside was a result of just such a powerful mechanism of change involving migration and adaptation.

During the late eighteenth century several thinking people had realized that these changes occurred. Erasmus Darwin had followed Blake into verse to express the web of complexity that was involved. The image was clearly in his sights even though he didn't know what to do about it, how to investigate it, how to experiment and test out theories on how it might work. All that had to wait for Erasmus's grandson and the time of scientific enquiry. For others during the Enlightenment it was enough to realize the existence of so many opposites and to begin to see how powerful they were. They showed up as wars between nations, hatred between religious groups and as very different forces in nature, some known, some unknown. It was competition between closely related species – grey squirrels displacing red squirrels, weeds diversifying in disturbed soils as part of the sequence of ecological succession.

These were the theories that Charles was to test against the whole challenging picture that the newly recognised scientists were bringing to biology, with empirical processes to match those already being used by chemists and physicists. In 1855 when he was getting down to absorbing the consequences of migration, and the impact of global geographical and climatic changes through deep geological time, Darwin stretched his new confidence against his fear. He wrote to Hooker that springtime: 'sometimes I think it will be good; at other times I really feel as much ashamed of myself as the author of the *Vestiges* ought to be of himself.'

Observations and experiments about migration

Life at Down had stimulated Charles to compare the relationships between human society and that in the rest of nature. The lives of London's citizens were vividly described in Dickens's novels and the political consequences were to last for over a century in the class-conscious society. How can the same process in other species kill off some and turn others more fortunate into different forms?

Darwin's breakthrough came with the realization that two processes were involved, not one: first, the selection of individuals that could adapt to the changes within a society, then, through different lengths of time, the divergence of this new group in a new environment. This was a simple view of how species and the environment formed an integrated part of the complex system that Darwin called 'an entangled bank, with many plants of many kinds, and birds singing on the bushes, with various insects flitting about'.

Darwin saw in that moment on the way home up to Downe how animals and plants are 'bound together by a web of complex relations'. 'From experiments I have tried,' he wrote, 'I have found that the visits of bees . . . are highly beneficial to the fertilization of our clovers. Humble bees alone visit the red clover as other bees cannot reach the nectar.' If one became extinct the other would follow. He told how a Mr H. Newman 'who has long attended to the habits of humble bees' believed 'that more than two thirds of [their nests] are destroyed [by field mice] all over England . . . Near villages and small towns I have found the nests of humble bees more numerous than elsewhere which I attribute to the number of cats that destroy the mice.' So cats controlled mice, and then bees, the abundance of red clover, and eventually the supply of honey. The number and kinds of species in the hedgerow were not there by chance but by some similar process of dependent interaction between several different species.

But how was this connected to migration? Darwin's observations from the *Beagle* showed that species could travel across vast ranges of geographical and climatic change and it was perhaps this that explained the enigma of migration. Darwin's first experience of island-hopping had been a big anticlimax. When the *Beagle* left Plymouth on 10 December 1831, the first part of the journey had given the inexperienced sailor days of wretched seasickness. They arrived at Tenerife on 6 January and, annoyingly, were forbidden entry to the port because of a suspicion of cholera on board. Captain FitzRoy had therefore decided to make straight for the Cape Verde islands.

Once there Darwin had noticed that the flora was very like that reported from mainland Africa. He was to see a similar connection on the Galapágos Islands where the plants and animals were like those on the mainland of Ecuador. These observations had inspired him to test his ideas that seeds could float across long distances at sea if they had sufficiently buoyant fruits to sail inside. At Down House he had been experimenting with different fruits and seeds, soaking them in sea water to show which species were most resilient and timed how long they took to sink.

Darwin's careful observations of the commonplace along the Sandwalk and in the garden at Down and in the fields of Kent and Sussex told him about the distribution of closely related species. He expected several species of the same genus to be located together, just as many of the Fagales were together in the copse and on the chalk escarpment beyond. Little did he realize that the fossil specimen of a deciduous leaf that he had collected from Tasmania in 1836 and then deposited in the British Museum was from the same Order of Fagales. He was not to realize that the group had migrated north up across the Americas and over to Europe: it was a journey that had taken millions of years and had begun around the time that the dinosaurs became extinct, 65 million years ago.

Other secrets were to be revealed from the evidence that the *Beagle* had brought back from South America, and Darwin checked with the specialists at the Zoological Society who had been looking at the specimens. Thomas Bell was one of Richard Owen's colleagues interested in vertebrates and had been puzzling over the iguana specimens sent from three of the Galapágos Islands. They were three distinct species. Specimens of mockingbirds had been sent as well and the ornithologist at the zoo, John Gould, also found that each island had distinct forms that had enough differences to warrant a unique species name. He named another new species of flightless bird *Rhea darwinii*; it came from a separate part of Patagonia, adjacent to the territory of a much larger species. The processes of divergence and species change showing up in these reptiles and birds from the other side of the world were the same as those that Darwin had seen with the hedge parsley.

In his 1844 essay Darwin had overlooked 'one problem of great importance' – the divergence of character. He didn't realize this until that journey in the carriage eleven years later. In his autobiography he recognized that moment and gave his outline of the issue:

> This problem is the tendency in organic beings descended from the same stock to diverge in character as they become modified. That they have diverged greatly is obvious from the manner in which species of all kinds can be classed under genera, genera under families, families under sub-orders and so forth; and I can remember the very spot in the road, whilst in my carriage, when to my joy the solution occurred to me; and this was long after I had come to Down. The solution, as I believe, is that the modified offspring of all dominant and increasing forms tend to become adapted to many and highly diversified places in the economy of nature.

But in his writings Darwin was reluctant to refer to details of the geographical separations within species or between them. He accepted that climate and physical conditions alone cannot explain different distributions. He wrote that distribution is often restricted within countries, that different species of the same genus were found close together and that competition from introduced species can overwhelm the native assemblages. He had not made up his mind where selection took place: was the choice between things inside cells, individuals, species or even higher ranking units?

In the middle of the nineteenth century no one knew anything about genes or their patterns of operation that we now call genetics. No one knew about plate tectonics, that continents of land on the earth's surface separate, move about and reconnect. Scientists were not sure whether parts of the web of nature split apart through divergence of groups. Another big difficulty was to understand how natural selection actually worked. What was it that was selected? What was the agent of selection? Was it in the individual organism, or a group of them that reacted to some kind of signal to develop as a new variety or even as a species?

There was so much uncertainty that it caused Victorian naturalists to be cautious in the way they lined up their evidence about evolution. Again, Darwin was not yet clear about the answers and for the moment said little.

'The Big Species Book'

In 1853 Darwin was awarded the Royal Medal by the Royal Society for his publications about his voyage on the *Beagle* and the systematic biology of the barnacles. The first sign that he was thinking about writing up his evolutionary theory is found in the entry in his pocket diary for 9 September 1854: 'Began sorting notes for Species theory.' Six months later he wrote to his friend William Fox:

> I am hard at work at my notes collecting and comparing them, in order in some two or three years to write a book with all the facts and arguments, which I can collect, *for and versus* the immutability of species.

It was 1856 and with his new ideas about divergence and with new confidence from the barnacles monograph as a job well done, Darwin was getting ready to start writing 'The Big Species Book', as his family at Down House were beginning to know it. By that year there were results from many of his breeding experiments and these helped with other sections within the book on the environment and common descent. There was plenty of inspiration coming from his own work and what others, such as Lyell, were reporting about more earthly changes from earthquakes and raised beaches. There were also plenty of alternative interpretations on the new evolutionary biology such as different opinions about vertebrate anatomy from Thomas Huxley and Owen. Also, there was the hope for a breakthrough in understanding how structural characters were transmitted from one generation to another and how breeding was able to sort out strengths from weaknesses.

Within two years, by the spring of 1858, the great manuscript was nearly finished. It was organized as eleven chapters that are familiar from the table of contents to the much shorter *Origin* that came out later. There were chapters on variation and adaptation, selection, hybridization and migration as well as one on the mental powers and instincts of animals. But instead of rising to the fame that this document deserved, it was put in a box in the broom cupboard under the stairs at Down House, with the parasols, croquet mallets and ancient, crooked tennis rackets.

Even worse, the whole manuscript was lost and forgotten for several decades, and when the box was eventually found, the dishevelled pages of handwritten manuscript couldn't be put into any useful sequence or context. It wasn't until the 1960s that the author's pocket diary covering the years 1838 to 1881 was recognized and this meant that the box of papers could be dated and put into the right order. It was a task done by their editor Robert Stauffer for publication by Cambridge University Press in 1975 as a volume called *Charles Darwin's Natural Selection: Being the Second Part of his Big Species Book Written from 1856 to 1858*. The book also provides references, footnotes and longer explanations of Darwin's ideas.

The manuscript was far from being a waste, however; it helped the author create the second shorter version with a much easier outline of the principles and with a clearer and easy style. Not only did 'The Big Species Book' contain a lot of long-winded stories, but in many places it showed confusion and wrongly worked-out parts of the arguments. There can be little doubt that Darwin left these uncertain parts until last. Without any knowledge of ecology and genetics he was unable to fully understand the nature of the evolutionary processes he was describing in the 'Big Species Book' and the direction they were taking him in. So it was just as well that he was forced to stop the project and start writing again in a different mode, leaving the contentious bits, and a lot of the detail, to one side. The papers from under the stairs nevertheless serve as a guide to how Darwin was thinking through the difficult mechanisms and logistics of everything involved in the big concept he called 'divergence'.

We can now see where Darwin was not clear in 1858, and understand where he had difficulty explaining what happened when the hedge parsley growing beside the Sandwalk first began to show signs of change. From what he had seen from his carriage window and elsewhere he assumed that the process of evolution began with some external environmental change. Proving that theory alone would need a lot of experimenting. Then there was the question of where the response mechanism in the hedge parsley was located internally, perhaps in some cells, and whether it was the cells within an individual plant only or a wider group of plants. In an extreme scenario, could it be that large numbers of individuals, whole patches of hedge parsley, all changed at once? Was this change perceived inside a cell or a tissue like a sensitive epidermis or an organ like a flower? Without genetics he couldn't know about the units of inheritance but he made some good guesses about the importance of sexual transmission and the importance of interbreeding. Nevertheless he had made up his mind that the two kinds of hedge parsley, one old and the other new, were battling out their territorial claims.

Darwin's argument in 'The Big Species Book' got into real trouble when he considered the interactions between these new neighbours, the reactions of the old ancestors and the new descendents and whether they separated or not. He tried to show how individuals within a species react to a new environment in four diagrams which showed the biggest changes at the edge of each tree. Still in awe of Lyell and the gradual rhythms of slow evolutionary change, extinction was bound to be difficult to explain. Darwin did not consider that catastrophes causing mass extinction events, or even much less dramatic environmental stress, brought urgency to the processes of evolution. In consequence, Darwin struggled: 'If the new species becomes isolated, or quickly adapted to a new station in which child and parent do not compete, both may continue to exist.'

Did extinction have a role within his emerging ideas? In order that descendents replace their ancestors, was extinction necessary? Although he was unable to come up with any clear answers Darwin continued trying to be objective about the processes he thought he saw in nature. Every day he looked out at the open landscape with the weather and ecology, insects and plants and the clues in the rocks about the geological past. He was the first to see life as an interactive web of cycles involving all these changing parts. He had a hunch that all living organisms are dependent on one another and on the whole environment and that they fit together in a functioning way.

Looking outwards and through geological time, on moving continents and with changing climates, he could see life push itself onwards. But there was so much going on at once and so little of it was even noticed let alone understood. While competition was highly regarded by Victorian society, little importance was given to understanding how something positive like the progression of species related to something negative like extinction. Instead, it was time to get results from some of the experiments going on in his garden.

6

Pigeons, Primroses and a Horse's Tooth

Seduced by the fanciers

On his rare trips to London Darwin often visited his pigeon club in Borough Market, close to London Bridge station. There he was known to the 'strange set of odd men' as 'the Squire'. Together they immersed themselves in talk of breeding strategies, sharing stories of their successes and failures. In 1856 Darwin wrote to his son William, who was away at school at Rugby:

> After dinner one of them offered me a clay pipe, saying 'Here is your pipe' as if it was a matter of course that I shd. smoke. – Another odd little man (N.B. all pigeon fanciers are little men I begin to think) showed me a wretched little Polish hen, which he said he would not sell for £50 & hoped to make £200 by her, as she had a black top-knot. I am going to bring a lot more pigeons back with me on Saturday, for it is a noble & majestic pursuit.

It also beat Darwin's own need to give attention to the hedge parsley, with its various shapes and sizes, and the different colours of its petals in different hedgerows. Whether it was the excitement of how the pigeon varieties would turn out, the fun of the alien pigeon fanciers' society in dingy London pubs, or the hope that these specimens would offer more clues to the evolutionary process, Darwin became increasingly fascinated by breeding his birds. He stopped his study of the migration of the hedge parsley and instead he moved his interest to another part of the garden at Down House. There, at least, he could write about his investigations at the same time as performing them.

'I have never met a Pigeon Fancier who did not believe in the evil of close interbreeding' wrote Darwin. To the enthusiast, inbreeding was a system used to concentrate desirable genes in a family. It showed up all the good qualities and it also showed up the faults. Some fanciers started out on an inbreeding program but gave it up due to the number of culls required. A cross between unrelated birds produced variability while inbreeding reduced the unpredictable randomness. Darwin wrote:

> I have found it very important associating with fanciers and breeders. For instance, I sat one evening in a gin-palace in the Borough amongst a set of pigeon-fanciers, – when it was hinted that Mr Bult had crossed his Powters with Runts to gain size; & if you had seen the solemn, the mysterious & awful shakes of the head which all the fanciers gave at his scandalous proceeding, you would have recognized how little crossing has had to do with improving breeds, & how dangerous for endless generations the process was. – All this has brought home far more vividly than by pages of mere statements & c.

Darwin observed that all these fashionable pigeon varieties that were discussed at the tavern table had evolved from the natural species. It gave him confidence that they were good experimental animals for his scientific work. He enjoyed taking time to decide which bird should be next for transfer from one cage to another. But he was soon to learn that it was as much an art as a science and that there was a very special breeders' culture as far removed from his middle-class world as it was from his scientific one. To his surprise, as well as that of his friends, he even enjoyed it:

> The other day [I] had a present of Trumpeters, Nuns & Turbits; & when last in London, I visited a jolly old Brewer, who keeps 300 or 400 most beautiful pigeons & he gave me a pair of pale brown, quite small German pouters: I am building a new house for my tumblers so as to fly them in summer.

His serious aim was to explore what he had learnt from barnacles in other groups like the vertebrates. He hoped to find universal trends in methods of reproduction, selecting the offspring to fathom how their evolutionary branch might look, and to see the slight changes in character from one species to another, or from one variety to another. For most of his ideas to be credible he needed actual evidence of what happened during breeding as well as in the resulting embryo.

Darwin had fond memories as a boy at home in Shrewsbury of his father breeding birds in pigeon lofts in the garden and many of his good friends were breeding the best specimens in their county; even Queen Victoria had a collection. William Tegetmeier, who lived in Wood Green in the north London suburbs, was one of the few middle-class experts breeding pigeons. He was a Fellow of the Zoological Society, poultry editor of *The Field*, and secretary of the Philoperisteron Society, a club for aspiring gentlemen and pigeon-fanciers. It met at the Freemason's Tavern in Holborn where its shows attracted hundreds of exhibitors. Tegetmeier had a good relationship with Darwin and they had regular correspondence about their interest. 'It would be better to cross some cocks & Hens of the half-breds from the two nests; so as not to cross *full* brother & sister,' Tegetmeier advised him

In overcrowded London pigeon-breeding was an opportunity to get away from the family and normal everyday stress. At Down House it was rather different. Darwin asked Isaac Laslett, the village carpenter, to build an aviary up against the orchard wall, complete with pens, protected from the bad weather. It was beside the greenhouse, the other indoor laboratory for the garden experiments, where he worked on his endeavour in both the animal and plant kingdoms. Here, Darwin performed the unpleasant task of boiling the pigeon skeletons with caustic soda, cleaning the bones so that he could measure the effects of the planned crosses of different breeds. The vertebrae and bones of the nose were especially important structures that showed differences between the varieties.

The skeletons revealed important clues that linked adapted behaviour to structure and Darwin wrote about many examples in 'The Big Species Book': 'Pigeon-fanciers select Pointers for length of body' due to the 'increasing number of vertebrae and breadth of ribs'. Encouraged by these results, he extended his observations to compare different breeds of dog. He found that 'with short-muzzled races of the dog certain histological changes in the basal elements of the bones arrest their development and shorten them and the position of molar teeth'. This link to development was another of his inspired suggestions that was going to take a lot more scientific study before its full value could be appreciated.

More results from plant breeding

Just outside the pigeon loft, and to the young gardener Henry Lettington's puzzlement, parts of the Kitchen Garden at Down were covered in fine netting. It kept insects off the primroses so there was no cross-pollination and the experiment allowed comparison with the cowslips that grew outside the netting. These experiments showed Darwin that while primrose and cowslip plants may have similar flowers they were definitely two distinct species. In recent articles in *The Gardeners' Chronicle* it had been stated that some primroses could suddenly change to this other type of *Primula*, the cowslip. This popular assumption would have undermined Darwin's belief that evolutionary change takes place gradually and so the Down House laboratory got busy. Throughout the spring of 1854 Darwin pollinated his primrose plants with the tip of his paintbrush, taking the pollen from one flower and gently rubbing it off on the style of another. He made these transfers between hundreds of flowers both on the same plant and on other plants in the greenhouse and then planted them out in the garden under the fine netting.

But the experiments provoked other questions. Did the fertilization of female parts on the same plant encourage inbreeding? Was sexual reproduction the key to good breeding? Was crossing unrelated individuals the most successful route? Darwin looked for clues about how flowering plants avoided the consequences, whatever they might be, by finding ways to breed out, to encourage cross-fertilization.

With such close-up work Darwin noticed that the primrose flowers in his garden were of two kinds: pin-eyed, with the style above the stamens; and thrum-eyed, with the style below the stamens. In his experiments Darwin discovered that for successful fertilization, pollen from a pin-eye plant must reach the style of a thrum-eyed one, or vice-versa. From these experiments Darwin was able to think about what evidence was needed to confirm that 'nature tells us in the most emphatic manner that she abhors self-fertilization'. The intricate reason for it, however, the advantages of resorting genes for each generation, was still elusive.

In the greenhouse at Down Darwin grew orchids, much to the pleasure of Betsy the parlour maid, who cut some of the flowers for the house, so that he could make observations of the flowers. His experiments showed how they made devious adaptations to attract bees in search of pollen. At this time he also wrote regularly for *The Gardeners' Chronicle* with hints based on some of the latest scientific discoveries. In one piece he set out an argument about the effect of cutting short the roots on the yield of fruit trees, and he offered an explanation that it depended on the concentration of particular chemicals. It was the earliest suggestion that there were such things as plant hormones. Another topic was how to avoid sterility that can come from some grafts.

Another puzzle requiring explanation was the familiar process of grafting fruit trees for different varieties of the same species or new hybrids. Darwin wrote a whole chapter in 'The Big Species Book' entitled 'Hybridism', writing about how easy it was to graft gooseberry on currant but how difficult it was the other way round. Why did different varieties of apricot and peach take the graft only on certain varieties of plum?

Yet all these experiments with pigeons, orchids, cowslips and primroses did not give him the answers he sought – how were characters passed from one individual to another? Despite this lack of progress in his breeding experiments he introduced the idea of pangenesis, his hypothesis that characters are passed somehow from one generation to the next by 'gemmules', anonymous agents of heredity. Gemmules were his idea of cellular particles produced by all cells that migrated from structural cells to sexual ones. They collected there to pass to the next generation at fertilization. Darwin never did find evidence to support the concept and was unable to grasp the notion that only some of the 'pangenes', as they were later called, were expressed, so enabling variation within individuals of the same species.

Vertebrates give up a few secrets

At least Darwin got great enjoyment from his pigeons, even though their value in elucidating some of his hypotheses was limited. A much more productive mid-nineteenth-century study of how vertebrates evolved came from the fossil bones being discovered by collectors in Europe and particularly in North America. In 1833 one of the most startling finds had been uncovered by Darwin himself while he was in La Plata, Argentina. In the same place he had also found skeletons from frightening mammoths like *Mastodon, Megatherium* and *Taxodon.* These woolly beasts with huge tusks roamed in southern climates after the last ice age and left plenty of fossil remains to fill the limited stowage space on the *Beagle* many times over.

But most exciting of all he had found the fossil tooth of a horse, easy enough to identify, but most difficult of all to explain. Although nothing was known of prehistoric mammals in that region, Darwin was puzzled because he knew horses had been absent from South America until the Spanish had introduced them in the fifteenth century. Was this fossil tooth evidence that they had roamed the pampas before this human intervention? If so, why had they died out, or had they just been hiding from human detection? More seriously, could the fossil prove Darwin's then very rudimentary theory of common descent for the vertebrates to be wrong? Had South American horses evolved outside the common group of all other mammals?

Darwin had taken the specimen to Richard Owen when he returned to London: 'Mr Owen and myself compared this tooth with a fragment of another, probably belonging to the Toxodon, which was embedded at the distance of only a few yards in the same earthy mass.' Owen was confident that the fossil was from an extinct species of horse but how he thought it came about was very different to what Darwin was thinking. According to Owen, the tooth was a fragment of the standard body plan that he had worked out from his French hero Cuvier. They expected that every living and extinct vertebrate species had limbs which were simple deviations from this ideal model.

Though Owen rejected natural selection he did support evolutionary change, though without a target of perfection he thought it an unnecessarily wasteful process. In 1850 he wrote to his sister Maria with a copy of the drawing of his essential plan: 'It represents the archetype or primal pattern, what Plato would have called the "divine idea" on which the frame of all vertebrates has been constructed.' Throughout his career Owen described and named hundreds of fossil and living vertebrate skeletons and continued to believe that each variation resulted from 'an idea in the divine mind which foreknew all its modifications'. It was a way to explain so much in nature, from catastrophe to headaches.

Owen's wife, Caroline, had more down-to-earth responses to his investigations, once protesting in her diary that 'the presence of a portion of a defunct elephant on the premises' made such a bad smell that she 'got R. to smoke cigars all over the house'.

in the late 1850s Owen worked on less aromatic specimens from the famous bird *Archaeopteryx* found in the Jurassic sediments of Germany. A little later he described bones of *Iguanodon* that had been found on the Weald just south of Downe by the Sussex solicitor Gideon Mantell. Knowing they were not lizards he assigned the fossils to a new group of reptiles that he named *Dinosauria*.

Owen believed that these different vertebrates had become extinct during Noah's flood, and conceivably earlier catastrophes, whose purpose was to make way for new species, and eventually, this process had led to the evolution of primates. It was the argument that FitzRoy had had with Darwin in his cabin on board the *Beagle*, looking out at the coastal plains of the Argentine pampas, where Darwin had found the horse's tooth. All three men agreed that the fossil was from an extinct species and that it fitted the explanation of an evolutionary lineage. It was no longer a case of the modern species having been killed off in South America by the deluge but a distant ancestor. This, in itself, raised difficult questions. Did evolution need the flood?

Owen was infuriated at the suggestion, especially since it advanced Darwin's reputation at Owen's expense. As Owen grew angrier about the reputation of the gentleman hiding away in the Kent countryside, Darwin found it easier to talk about the evolution of horses with his friend Thomas Huxley, one of the Royal Society's bright young fellows. Before he became involved with the evolution of mammals Huxley had specialized in marine invertebrates and by the age of thirty-five he had coined the very different words Hydrozoa and agnostic. These topics showed the diversity of his interests and that he had a place in science for open-minded creativity. Hooker had introduced him to Darwin on his return from the southern hemisphere as HMS *Rattlesnake*'s surgeon. Huxley was to become a strong defender of the new thinking about natural selection.

With newly discovered fossil bones from Europe, in the 1840s Huxley and Owen had set about reconstructing the actual limbs of three species of horse. To disprove any role of a global flood and to show that evolutionary pathways are made up of extinct descendents was a double challenge. The oldest specimen was the four-toed *Eohippus* which was no bigger than a dog, then the larger three-toed intermediate species, and thirdly a species from an interglacial deposit less than a million years old, an extinct species of the big one-toed familiar *Equus*. It was just the right kind of illustration of an evolving lineage for Huxley to describe in his guest of honour lecture at the forthcoming opening of Johns Hopkins University in New York.

Huxley had constructed a tentative evolutionary lineage based on the European specimens. He did not realize they were just the edge of a big evolutionary tangle that involved migration and the physical drift of entire continents, and depended on the concurrent evolution of very different species. To make things worse, his host in America, Othniel C. Marsh, had sorted out the collection of extinct horses in North America, and showed him yet another evolutionary pathway. It comprised a straight lineage from one horse species to another, four toes to three to one, all species leading up to bigger and better forearms, legs and teeth. Marsh had assembled new collections of extinct horse skeletons from the Tertiary sediments of Wyoming and Colorado that overwhelmed Huxley and Darwin. It was immediately clear to them that the three European species had migrated eastwards to Europe over the last fifty million years. But how did these land mammals get there?

Darwin also talked with Hooker, just back from Calcutta, about Indian elephants but the poor fossil record led them nowhere. They discussed the likely shape of the genealogy between the extinct species but there was insufficient evidence to reconstruct even a hypothetical evolutionary tree, even though they looked north and south for more examples of that and other branches of the vertebrates. Darwin had thought a little earlier that the Indian elephant might show up a good history of skeletal change and he was convinced that the species *Elephas indicus* contained two breeds that he might prove to be distinct species. One was thick-set, courageous, with short tusks directed inwards so that when attacked by a tiger it fell on its knees and used them to pin the tiger to the ground. The other couldn't do this because it had a bigger head.

Remembering the hedge parsley

One day around this time, Darwin went off as usual on his old horse Tommy who 'stumbled and fell rolling on him and bruising him seriously'. Etty wrote to George:

> We've had a very unpleasant event this week. The immaculate Tommy has thrown Father. They were cantering over Keston Common when Tommy tripped & fell bang down – so completely head over heels that his ears & the pommel of his saddle were the two parts muddied. Father of course imitated Tommy's movements . . . someone soon came to his help & he was taken into a house & lay down on a sofa for a bit. After ½ an hour Tommy was caught & as the fly Father ordered was very long in coming he got on Tommy and was led home. I fear another Tommy will never be found & I fear Father's nerve will be considerably shaken so it is altogether a bad job.

The physical pain added to his worry about the storm he expected after publication of his ideas.

Pacing along the Sandwalk, counting the times he walked around the copse, Darwin soon returned his thoughts towards the hedge parsley, and the importance of its separation through space and time. The horse crossing the world was another example in vast scale of what was occurring in the garden at Down House – the adaptation and migration of the plant. But if the horse had migrated over such large distances it also must mean that other forms had made that journey. How had the horse travelled? Was the world once a single land mass? Did one particular catastrophe force the animal to migrate? Was this a history that was shared by humans? Darwin hoped that by returning to the Sandwalk he would gain some answers.

On one hand, evolutionary change could be explained by Lyell's gradualism, and on the other by the South American catastrophes. But neither explained why the hedge parsley should have changed. Was there some internal cellular catastrophe?

Another thought occurred to Darwin on his circuits around the copse. Unlike the hundreds of living species closely related to the hedge parsley, the examples of vertebrates that he was contemplating were atypical because both *Equus* and *Homo* had only one surviving species from a colourful diversity in the fossil record. Now we know that just like many other species, humans also had their own breeds, or races, and their own migratory history, out of Africa, across the Levant, Neanderthals splitting off in western Europe and becoming extinct in recent geological time. These single species were the outcome of a wide range of forms that adapted and overcame the changes that made their competitors extinct.

The ancestral and geographical origins of two other big groups of vertebrates from the southern hemisphere also gained Darwin's attention. Reports were arriving back in England of the discovery of new species of marsupials and flightless birds, all restricted to this region, though their fossil record extended northwards up into South America. The new evidence fitted Darwin's thinking about divergence and common ancestry. Away from competitors the marsupials and flightless birds could survive. Imposing this evidence from a grand global scale of observation upon the smaller scale experiments in the garden allowed Darwin to see a vast web of evolution while filling in many of the details about what went on between the organisms themselves, their anatomy and their organization. Bringing these two ends of the argument together, each part could answer the questions raised by the others.

Life at Down House was becoming busy and all the while he carried on breeding new varieties of pigeon, counting the numbers and proportions of pin- and thrum-eyed primroses. The garden made more conventional demands as did the family and his growing number of correspondents. But Darwin's inventiveness was being challenged to find evidence for two of the main outstanding issues, the mechanisms of inheritance and divergence. Try as he did to detect the gemmules that he proposed as the agents of heredity, all signs remained elusive. What was more, the ecological and geological stimuli that were driving divergence also continued to avoid detection. The frustrations came to the fore one morning in that summer of 1858 when a catastrophic event disturbed the peace at Down House.

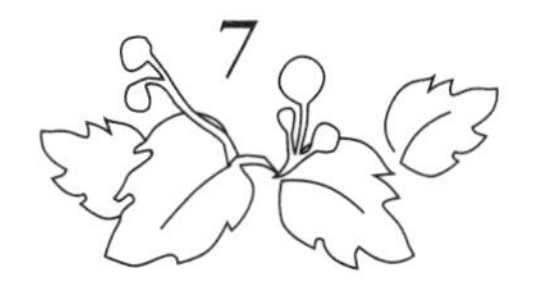

7

Actions out of Quietness

On 18 June 1858 the postman called at Down House with a package sent from Ternate, an island close to New Guinea. It was from someone else who, so it turned out later, had also just read Thomas Malthus's warnings about the limits to the survival of human population. The arrival of the package was well timed for Darwin needed to have a catastrophic jolt himself, and this one was to change his life. The shock was made worse by his own isolation at Down House, away from support of the latest talk in London society.

Alfred Wallace was a professional collector of tropical species. Three years earlier he had addressed another letter to Down House with a copy of his 1855 article that had been published in the *Annals and Magazine of Natural History*. In it Wallace argued that the relationships between closely allied species showed powerful geographical patterns and hinted that these pressures might generate some evolutionary change. Darwin had not noticed the innuendo and in his congratulatory reply to Wallace he wrote asking for specimens and observations from the Malaysian rain forests. His letter mentioned nothing about evolution.

The latest package contained another manuscript written by Wallace revealing the importance he also attached to Malthus's arguments about population growth:

> Disease, accidents, war and famine would keep down the population of savage races. This self-acting process would necessarily improve the race because in every generation the inferior would inevitably be killed off.

Reading these words Darwin was overwhelmed with fear that Wallace was going to steal his thunder and beat him to the post with this early publication. But the manuscript was for Darwin to pass on to Lyell for comments, not yet to be offered to a journal. Nevertheless, it meant there was no time for the sage of Down House to finish 'The Big Species Book' before Wallace's manuscript would be published.

Hastily Darwin returned to his study and wrote to Lyell:

> [Wallace] has to-day sent me the enclosed and asked me to forward it to you. It seems to me well worth reading. Your words have come true with a vengeance – that I should be forestalled. You said this, when I explained to you here very briefly my views of 'Natural Selection' depending on the struggle for existence. I never saw a more striking coincidence; if Wallace had my MS Sketch written out in 1842, he could not have made a better short abstract! Even his terms now stand as heads of my chapters.

A week later, on 25 June, Darwin wrote another letter to Lyell asking for his advice.

> I should be extremely glad now to publish a sketch of my general views in about a dozen pages or so; but I cannot persuade myself that I can do so honourably. Wallace says nothing about publication, and I enclose his letter. But as I had not intended to publish any sketch, can I do so honourably, because Wallace has sent me an outline of his doctrine? I would far rather burn my whole book, than that he or any other man should think that I had behaved in a paltry spirit.

Back from botanizing in India to take the position of Assistant Director to his father at Kew Gardens, Joseph Hooker had just become a leading fellow at the Linnean Society and with Lyell's support saw to it that both the paper Wallace had sent in the package and a hastily prepared article by Darwin would be read to a meeting of the society as soon as possible. Still overwhelmed with surprise that someone else had had the same ideas, Darwin was only too pleased to put the matter of presentation into the hands of his two friends. The two papers were pushed on to the programme of July's meeting, read in the authors' absence and received 'no semblance of discussion' from the twenty-five members of the audience. They had turned up to pay their respects to the lately deceased botanist Robert Brown and weren't interested in the surprise additions to the agenda.

The procedure gave licence, however, for the two papers to be published in the society's journal, in which they appeared in August 1858. Wallace had known nothing of the presentations to the society and Darwin wrote carefully and apologetically, afraid of how he would react. He began the letter politely and set out his plans for a 500-page abstract. Then he referred to their joint paper to the *Journal of the Linnean Society*:

> Though I had absolutely nothing whatever to do in leading Lyell and Hooker to what they thought a fair course of action, yet I naturally could not but feel anxious to hear what your impression would be. I owe directly much to you and them.

In the second paragraph of the letter he went on to other more important matters: 'I am glad to hear that you have been attending to birds' nests.'

On 13 July Darwin wrote to Hooker thanking him:

> Your letter to Wallace seems to me perfect, quite clear and most courteous. I do not think it could possibly be improved, and I have today forwarded it with a letter of my own. I always thought it very possible that I might be forestalled, but

> I fancied that I had a grand enough soul not to care; but I found myself mistaken and punished; I had, however, quite resigned myself, and had written half a letter to Wallace to give up all priority to him, and should certainly not have changed had it not been for Lyell's and your quite extraordinary kindness. I assure you I feel it, and shall not forget it. I am *more* than satisfied at what took place at the Linnean Society.

And then, putting his priority back into place, Darwin ends with a postscript:

> I have had some fun here in watching a slave-making ant; for I could not help rather doubting the wonderful stories, but I have now seen a defeated marauding party, and I have seen a migration from one nest to another of the slave-makers, carrying their slaves in their mouths!

On 18 July he wrote to thank Lyell with a different order of priorities:

> I have never half thanked you for all the extraordinary trouble and kindness you showed me about Wallace's affair. Hooker told me what was done at the Linnean Society, and I am far more than satisfied, and I do not think that Wallace can think my conduct unfair in allowing you and Hooker to do whatever you thought fair. I certainly was a little annoyed to lose all priority, but had resigned myself to my fate.

At the same time as these traumatic events another catastrophe happened in the Darwin household. Emma and Charles's two-year-old son, Charles, contracted the epidemic scarlet fever and died on 28 June. Yet again, the child's parents were brought back to the tragedies of their own parents, the deaths of Mary and Annie, and their anxiety about inbreeding. Fortunately they both had enough fortitude to get on with their rich life at Down and Charles had another book to write, quickly. It was to be much shorter than 'The Big Species Book' and more in the style of his literary friends. He explained the endless variations of nature with the same basic set of patterns, seizing on small points to make one large one.

Writing began in August and the author soon realized that there was plenty of work remaining. In October he explained to Hooker:

> I am working most steadily at my Abstract [*The Origin of Species*], but it grows to an inordinate length; yet fully to make my view clear I cannot make it shorter. It will yet take me three or four months; so slow do I work, though never idle. You cannot imagine what a service you have done me in making me make this Abstract; for though I thought I had got all clear, it has clarified my brains very much, by making me weigh the relative importance of the several elements.

The manuscript was ready to go to John Murray the publisher the following May, 1859. Its contents were based on the same chapter headings as those he had used in 'The Big Species Book', but it was without sources, footnotes or references. *On the Origin of Species by Means of Natural Selection or The Preservation of Favoured Races in the Struggle for Life* was published in November 1859. The first edition was bound in green cloth, was 502 pages long and cost fourteen shillings (70 pence). A total of 1,250 copies were printed and Darwin wrote to thank the publisher: 'I have received your kind note and the copy; I am infinitely pleased and proud at the appearance of my child.' The author was worried that he had made so many changes to the proofs and thanked Murray for bearing the cost of putting them right, £72 8*s*. It was a small price to pay for a first edition with no printing errors whatsoever.

Publication of the *Origin*

One of the first letters of congratulation, also thanking him for his copy of the *Origin*, came unexpectedly from the pen of Richard Owen who praised the 'continuously operative creating forces' to explain biodiversity. Darwin was pleased with the letter and genuinely wanted to know more of what Owen really thought, so he decided that when he was next up in London he would call round and see his friendly adversary. Owen was an influential scientist and Darwin worried about what effect his forthcoming review of the *Origin* would have on the likely readership.

On the few occasions that he was now in London, Darwin had become a reluctant and cautious participant in the social round. It was a serious handicap as calling on acquaintances was how people communicated and did business, being greeted by a servant at the main entrance to the house, handing over a visiting card, waiting nervously for the host to appear. There was plenty for the two scientists to discuss, if only they could put their different attitudes about religion and politics to one side and bury their prejudices about the French.

Darwin should have known better than to call in this way and the timing of the visit was bad for other reasons that Darwin hadn't appreciated. At that time, as a curator at the British Museum in Bloomsbury, Owen was trying to obtain funding from the government to establish a separate museum to glorify God's handiwork in nature. He wanted a new building in South Kensington and for it to be a Creationist shrine, to hold the dinosaur skeletons that had recently been at the Great Exhibition near by. These remains had also been the basis of the models of extinct reptiles he exhibited at Crystal Palace. Hooker had opposed the idea, as did Darwin, both arguing that specialist collections shouldn't be separated from the rest of human endeavour and campaigning for the specimens to stay at Bloomsbury. They were also worried that the Creationist arguments Owen wanted to associate with the specimens might give the public an extreme view of how life diversified. Owen's opinions on the shape of evolution were questionable and the opposition became personal when Lyell left Owen's name off a list of Europe's greatest biologists. Darwin's book only made the angry anatomist even more jealous of Darwin with his new book.

Unaware of these hurts, Charles Darwin knocked on the door of the Owens' house in Bloomsbury and gave the maid his visiting card. He was suddenly wondering what to say, confused about his priorities if asked to defend his theory. But Richard Owen was less interested in Darwin's new book than in his own ideas and his reputation as an anatomist and a Creationist. He said as much to his shy visitor who listened politely, passed the time of day and left. In a town full of this kind of politicking Charles Darwin wanted to do just one thing, to go home to his garden at Down House.

Three months later, when his friends Charles Lyell, Thomas Huxley and Joseph Hooker were staying, the March 1860 edition of the *Edinburgh Review* was pushed through the letter-box. It was just as well that all three of them were together because the magazine contained a 45-page review of *The Origin of Species* that was highly critical of the work of all three of them and their 'vague and general' approach to evolution. As was the custom in those years, the article was written anonymously but it soon became clear to them that the author was none other than Richard Owen. He was one of Britain's most controversial biologists so they had been expecting to hear his reaction, but what they read that morning by the fireside at Down House was more vicious than any of them had expected.

'Professor' Owen's review flattered 'Mr' Darwin's style and powers of observation but showed no sympathy for the self-control of natural selection. Mischievously, Owen kept referring to his alternative theory of archetype and homology, quoting his own work on vertebrates and pushing his idea of 'a continuous creative cooperation of the ordained becoming of living things'. This was Owen's Law of Irrelative Repetition, the multiplication of organs performing the same function, like worm segments and vertebrae. He argued that this was the method used by God to create diversity. It was based on the common beliefs of the older generation, the Naturphilosophen of Goethe in Germany and the divine homology of Cuvier in France. Owen put them together to account for the serial differences within an organism from top to bottom – the leaves and branches on a stem, the limbs and protrusions from a backbone.

Owen's review was like a red rag to a bull for Thomas Huxley, who was gaining his own experience of vertebrate anatomy in his new job as a palaeontologist at the Royal School of Mines, now part of Imperial College. With this group of supporters now united by Darwin's book, the opposing conservative establishment was able to focus their argument. They were worried that if they gave way on this principle of natural selection, then threats to their political position and all that went with it would follow. The battle was an important part of the intellectual war between science and religion and Richard Owen was a fine general to lead the defence.

The best-known reaction to the publication of *The Origin of Species* came at the 1860 meeting of the British Association for the Advancement of Science, that year held in Oxford, and hosted by Bishop 'Soapy Sam' Wilberforce, a short and fat man, whom Huxley later described as 'oily with a round mouth, and spoke as though he had a plum in it'. Dickens must have known him as the model for many characters but Richard Owen both knew and got on with Sam and was invited to stay in his house overnight before the meeting's big lecture about *The Origin of Species*. Unaware of the arguments, Sam needed briefing before he chaired the session and Owen generously provided him with notes for the meeting.

Between 700 and 1,000 people turned up to be addressed by Professor John Draper from New York University on 'Darwin and Social Progress'. The audience was too big to fit into the room that was allocated so they had to move into the library. Hooker wrote a report of the meeting to Darwin, who never attended affairs like this, saying that Draper

> droned on for an hour, no one went out, indeed no one stirred. It was a pie of Herb Spencer without the reasoning. Sam Oxen [Bishop Wilberforce] got up and spouted for half an hour with inimitable spirit, ugliness and emptiness and unfairness. He ridiculed you badly and Huxley savagely.

Then the bishop turned to Huxley. 'Was it on your grandfather's side or grandmother's side that you were descended from an ape?' According to Francis Darwin's *Life and Letters*, 'Huxley replied to the scientific argument of his opponent with force and eloquence, and to the personal allusion with a self-restraint, that gave dignity to his crushing rejoinder.'

Many versions of Mr Huxley's speech were current: the following report of his conclusion is from a letter addressed by the late John Richard Green, then an undergraduate, to a fellow student, now Professor Boyd Dawkins:

> I asserted, and I repeat, that a man has no reason to be ashamed of having an ape for his grandfather. If there were an ancestor whom I should feel shame in recalling, it would be a *man*, a man of restless and versatile intellect, who, not content with an equivocal success in his own sphere of activity, plunges into scientific questions with which he has no real acquaintance, only to obscure them by an aimless rhetoric, and distract the attention of his hearers from the real point at issue by eloquent digressions, and skilled appeals to religious prejudice.

The meeting quickly turned to farce: a tightly corseted lady fainted; the hook-nosed FitzRoy, former captain of the *Beagle* and now head of the government's meteorological department, then stood up and, Hooker reported, 'lifting an immense Bible first with both and afterwards with one hand over his head, solemnly implored the audience to believe in God rather than man'. The crowd shouted him down and the meeting ended in uproar. Two days later Darwin wrote to Hooker, 'I would have soon have died as try to answer the Bishop in such an assembly.'

The event was reported in the newspapers and drawing-rooms of the land and soon became apocryphal. Four years later the Tory Prime Minister Benjamin Disraeli addressed the Oxford Diocesan Society, wearing a black velvet shooting-jacket and a Robin Hood hat. He announced to Wilberforce: 'The question is this. Is man an ape or an angel? My Lord, I am on the side of the Angels.'

The three men who read the reviews of the *Origin* around the fireplace at Down House were scientific pioneers and they knew they had to break from the old practices. Owen advocated his own sense of purpose, offering wise guesses and then testing each hypothesis. Huxley, Hooker and Darwin, on the other hand, wanted to go a stage further and separate all scientific study from any religious orthodoxy. The three friends together in Down House in March 1860 were thereafter reluctantly bent on a philosophical and political mission.

Yet even within this group there were different attitudes towards evolution and different senses of perspective in looking at it. Huxley was thinking of an organ suddenly adapting from one form to another, what became known as a mutation. For some time he had also regarded natural selection as the 'gladiatorial theory' of existence, even seeing sexual posturing and cooperation as combative traps leading to the deaths of the losers. Nature, he argued persuasively, had a habit of acting in an opposite manner.

The third man, Joseph Hooker, has often been accused of never really understanding what Darwin had in mind for natural selection. Instead he just wanted to be able to give names to different specimens so that plants could be fixed components in his identification of whole floras. The lists of species that he had compiled from New Zealand were different to those he had from India and he wanted to describe them unambiguously. At subsequent meetings together the three men would argue about whether evolutionary change was gradual or sudden and whether some species might even be fixed, with no evolution at all for long periods of geological time. These were big issues that troubled Darwin a lot and that winter early in 1860 he was planning experiments in his garden to investigate them.

The fourth man who should have been at Down House that day was Charles Lyell. It took slightly longer for Darwin to hear his response and it turned out to be greatly disappointing. Lyell's feelings on the matter are explained cryptically in his diary for March 1860 where he compared evolution to the 'Hindoo Triad'. He could easily follow how it involved Vishnu the preserver and Shiva the destroyer. What he could not understand was how it was also composed of a force working in the role of Brahma, the creator. No one could give him a clear answer, not even Darwin, and their relationship cooled as their opinions hardened.

In 1863 Lyell responded to the *Origin* by publishing his own manifesto, *The Antiquity of Man*. Although he introduced the idea of missing links, in presenting the geological history of man Lyell was still arguing for an unknown gulf between man and beast. Darwin was devastated, his feelings expressed meekly by scrawling in the margin of his copy of the book 'Oh'. Later he admitted:

> The Lyells are coming here on Sunday Evening to stay till Wednesday. I dread it, but I must say how much disappointed I am that he has not spoken of a Species, still less of Man. And the best of the joke is that he thinks he has acted with the courage of a Martyr of old.

The visit was soon cancelled by Emma on account of her husband's over-sensitive stomach.

Another review of one of the later editions of the *Origin* drew attention to the first four paragraphs of the introduction, where the words 'I,' 'me,' 'my,' occurred forty-three times! In one of his many letters to Hooker that year, 1863, Darwin joked that he

> was dimly conscious of the accursed fact. [The reviewer] says it can be explained phrenologically, which I suppose civilly means, that I am the most egotistically self-sufficient man alive; perhaps so. I wonder whether he will print this.

Then in a postscript: 'Do not spread this pleasing joke; it is rather too biting.'

Celebrations with the X Club

One evening in November 1864, just after a meeting at the Royal Society, then in Burlington House on Piccadilly, a group of five fellows walked round the corner to the St George's Hotel in Albemarle Street for a drink. They included the Irish physicist John Tyndall, Herbert Spencer, Darwin's cousin the polymath Francis

Galton, Joseph Hooker and Thomas Huxley. Eventually there were nine members of the group and to show they were expecting to grow with one more they called themselves the X Club.

This was the founding meeting of a dining club in defence of evolutionary naturalism which was to become a powerful pressure group with strong influence inside the Royal Society as well as government and other British institutions. They were to organize lectures, articles in the newspapers and appointments, and campaign for the end to the abuse of science in the new industries, much in the spirit of the old Lunar Society. But most especially this group was to propagate the meaning and scope of evolution by adaptation and natural selection.

Darwin was never recorded as a member of the X Club but several meetings with his closest friends Hooker and Huxley at Down House suggest that he was its tenth member. There is no evidence that he attended their meetings, or their celebrations, though they did meet informally at Down. There, the three friends would look out from the drawing-room on to the garden through the wooden veranda, ivy and other plants covering the pillars. They would walk beyond the flower beds planted out with Emma's favourite selection of perennials and shrubs, primroses and cyclamen and fashionable Dorothy Perkins roses. A place to sit was under the mulberry tree, while further away was the shade of the big sweet chestnut, a Scots pine and yews. There was the sundial that they used to check their watches casting a shadow on the closely cut lawn where they would play croquet. They would see Darwin's cucumbers, whose tendrils he found 'coiling to the touch', and some of the experiments that were later described in his 1865 essay 'On the Movements and Habits of Climbing Plants' published in the *Journal of the Linnean Society*. In Darwin's company these were the influences that the X Club members experienced. Political campaigns were left for London and his absence.

Gradual or sudden change

It was not until the publication of an edited version of the unfinished manuscript of 'The Big Species Book' more than a century later, in 1975, that there was any idea that Darwin had been seriously considering an alternative to his ideas of gradual evolution, with hints of catastrophic change from sudden events. Did this uncertainty stay in his mind as a range of options or was the popular theory of gradual change enough to explain his experiments? He had noticed changes in the number of segments in the antennae of a beetle species and the shapes and sizes of evening primrose variants. The different forms appeared suddenly and were stable in future generations, suggesting that sometimes evolution could go in jumps. The observations no doubt caused him many sleepless nights and stomach cramps, but he had no clear answers. Darwin never went on to support either extreme on the spectrum from gradual to sudden rates of evolution. Maybe there was no decision to be made at all, for within such a very big network as nature all the routes, and others in between, may become operational at some time or another.

The web of nature was made up of different physical shapes and substances, from volcanoes to molecules, some rare, others common, some changing slowly while others transformed quickly. There was no telling whether environmental changes might be slow or quick and Darwin had experienced both extremes and plenty in between. But his reputation was based on a strong image of a slow and bearded old man advocating the power of small change over big time. His long interest in earthworms, on which he conducted experiments over many years, showed that they sorted the earth grain by grain to give the best soil and topography. So he became quite indignant when a Mr Fish denied that worms account for much 'considering their weakness and their size'. Darwin responded: 'Here we have an inability to sum up the effects of a continually recurrent cause, which has often restarted the progress of science.' Needing to summarize his contributions to biology, at the end of the *Origin* he returned to the same theme:

> We are always slow at admitting any great change of which we do not see the immediate steps. [Natural selection] is daily

> and hourly scrutinizing throughout the world, every variation, even the slightest; rejecting that which is bad, preserving and adding up all that is good; silently and insensibly working.

As well as having 'worms crawling through the damp earth' of his famous entangled bank Darwin also had 'various insects flitting about . . . dependent on each other in so complex a manner'. It was what he and Emma had seen in the summer just after the *Origin* was published when they were walking through the Ashdown Forest 10 miles (16 kilometres) south of their home. There in the acid soil was the small, very rare bog orchid and next to it the sundew, also dependent on insects but for food not pollination. These insectivorous plants were cultivated for testing the ways in which they attract insects and how the glandular leaves bend over to trap them and eventually devour them, and so Darwin collected some and brought them back to Down for experimenting on. Emma was by now resigned to their country walks continuing to be interrupted and turned into collecting expeditions to stock the greenhouse for more experiments.

Yet Darwin's tireless search for the two other pieces of evidence missing from the *Origin*, which logic told him to expect, went on. One was for the unit of inheritance that was transmitted through sexual reproduction, what he had called the gemmule, now called the gene. The other was for a motive or some kind of driving force to evolution. He had a suspicion that some cellular processes might stimulate developmental variation of form and he wondered whether science would some day find all the answers.

8

Exploring the Gradual

Séance or science

Imagine what a fuss the media would make of it today: a séance of twenty famous people conducted by a leading spiritualist well known for attempting to communicate with the dead. These experiences with the afterlife were, however, common social occasions in the later part of the nineteenth century. They helped some people cling on to the declining trust in God as their society was entering an age of new influences from science and technology. Part of the ritual had the participants checking that nothing was hidden behind the curtains or under the carpets. On this occasion in 1871 the sceptical host was Erasmus Darwin, Charles's older brother, who had been cajoled into inviting a group of participants with widely different views about the spectacle.

Surprisingly, several members of the X Club attended that day. There was Galton, Hooker, Wallace and even Huxley, though he insisted on being disguised so that he was not recognized outside or even inside Erasmus's house. While they were waiting for the proceedings to begin Emma and Charles talked to Mary Anne Evans, who wrote under the pseudonym George Eliot, about her new novel *Middlemarch* that was receiving very good reviews at the time. It was another story, set in the 1830s, of the everyday consequences of the social reform and scientific advances that were taking place in England. The Darwins were curious whether these same trends had brought them all together at that strange gathering.

The question was soon answered as three of the participants reacted very differently to the event, each revealing more about their own social position and personality than the influence of any spirits. One was Charles, who walked out after half an hour complaining that it was too hot and claustrophobic, but really he was very angry that such an unscrupulous performance should be taken so seriously. Huxley stayed to observe and went on later to expose the spiritualist leader as a cheat and impostor: 'I never cared for gossip in my life; and disembodied gossip such as these worthy ghosts supply their friends with, is not more interesting to me than any other.'

Wallace also stayed on and took it all in. A few years earlier, in an article written for the *Quarterly Review,* he had shown interest in what he called spiritualist forces acting between humans. He had written that natural selection was the mechanism by which nature derived such a range of diversity in its animals and plants, but it could not explain human consciousness or how man felt in his soul. Humans had some higher driving force so that 'the true grandeur and dignity of man is . . . a being apart, since he is not influenced by the great laws which irresistibly modify all others'. In his own copy of the magazine Darwin wrote in the margin 'No!!!' and underlined it three times. Then, pleading Wallace to think again, he wrote a letter:

> I hope you have not murdered too completely your own and my child . . . If you had not told me, I should have thought that [your remarks] had been added by someone else . . . I differ grievously from you, and I am very sorry for it.

Wallace was sure that the metaphor of a Tree of Life was a good way to describe the shape of biological evolution. In another of his essays, about his travels on the Sarawak river, he had linked the diverging pattern of common ancestry to plant geography: 'When a group is confined to one district, and is rich in species, it is almost invariably the case that the most closely allied species

The rear elevation of Down House, facing south-west. A mulberry tree is in front of the right-hand part of the building.

The greenhouse beside the south-facing wall of the Kitchen Garden. The orchard is over the wall, as is the brick potting shed that Darwin had built near the end of his life.

The Sandwalk, facing south towards the summerhouse, with the copse on the left and Great Pucklands Meadow on the right.

Oil painting of Dr Erasmus Darwin (1731–1802) by Joseph Wright of Derby, about 1770.

Robert Darwin (1766–1848). A physician like his father Erasmus, Robert married childhood friend Susannah Wedgwood in 1796. He was strongly in favour of his son Charles following the family tradition and studying medicine, but eventually supported Charles's decision to pursue his interest in natural history.

Like the rest of Down House, the study was sparsely furnished. It was here that Darwin wrote *The Origin of Species*.

Charles Darwin (with paper in hand), Joseph Hooker and Charles Lyell (l to r) in the study at Down House. The two scientists were frequent weekend visitors to the Darwin family home.

(right) An elderly Darwin experimenting in the potting shed at Down House. Darwin continued his experiments long after the publication of *The Origin of Species* in 1859.

(left) Alfred Wallace started his career as a school teacher, became a professional field naturalist and shared Darwin's announcement on the theory of natural selection at the Linnean Society.

Thomas Huxley (1825–1895). Huxley was a staunch supporter of Darwin's theory of eveloution

Darwin's cousin, Francis Galton (1822–1911).

Edward Forbes (1815–1854) was a bright young marine biologist and a popular house guest at Down House.

William Bateson (1864–1926), the first man to use the word 'genetics' and the founder of the *Journal of Genetics* in 1910. He eventually became a professor at Cambridge University.

(left) Richard Owen (1804–1892) became a leading critic of Darwin's ideas about natural selection. Controversially, he moved the natural history collections from the British Museum in Bloomsbury into the purpose built Natural History Museum at South Kensington.

Charles Darwin on the veranda at Down House, about 1880.

are found in the same locality, and that therefore the natural sequence of the species by affinity is also geographical.' But unlike Darwin, Wallace was clear that human beings were at the top of the Tree, confident in their supremacy over all other nature, while Darwin didn't seem to care about such priority. For him, any place for humans or any other species was going to be ambiguous.

There were some features of evolutionary biology that were becoming clear to Darwin and an ordinary place for humans on the Tree of Life was one. As he became more knowledgeable and self-assured, he also began to challenge Lyell's influence, especially about gradualism. So he kept his own mind open to as many possibilities as possible. He was, after all, inspired by two extremes: the effects of the earthquake he had witnessed in Chile and the quietness in which he lived at Downe.

The diverging opinions of the men at the séance were just another example of the divisions facing many thinking people caught between the traditional assumptions of the day and the reforming dreams of the future. Although most members of the X Club followed Darwin, Wallace couldn't quite abandon all his theological beliefs concerning questions of the soul. It was a strong view with much support, that humans were still the proud climax of evolving diversity, and it followed that science was the latest tool that would prove the point. Others, including Darwin, thought that it followed logically from the theory of common descent that *Homo sapiens* was part of the single primate branch of mammals and that there was no need to treat our species differently to any of the others.

This debate was linked to the question of whether humans could control nature. Man certainly had changed the environment much more than any other species had ever done, but can man control what's happening on the planet? Science was regarded by many to be on the way to answering questions about rising populations and the creation of life. Earlier, some biologists had followed Linnaeus and called the unknown powers that provided the driving force to nature, the 'essence'; later, in the early twentieth century, the French biologist and philosopher Henri Bergson called it 'l'élan vital', the vital impulse.

Mesmerism was one fashionable practice that some thought might help reveal many of these processes. But these and similar unspecific concepts that were often talked about then, such as the forces between living beings, and the controlling power of nature, were hard to relate to scientific treatment. Darwin had entertained the possibility of their importance but didn't know how to proceed and he had written to his old friend at Cambridge, William Fox:

> With respect to mesmerism, the whole country resounds with wonderful facts or tales. I have just heard of a child, three or four years old (whose parents and self I well knew), mesmerized by his father, which is the first fact which has staggered me. I shall not believe fully till I see or hear from good evidence of animals (as has been stated is possible) not drugged, being put in stupor; of course the impossibility would not prove mesmerism false; but it is the only clear *experimentum cruces*, and I am astonished it has not been systematically tried. If mesmerism was investigated, like a science, this could not have been left till the present day to be done satisfactorily, as it has been I believe left. Keep some cats yourself, and do get some mesmerizer to attempt it. One man told me he has succeeded, but his experiments were most vague, as was likely from a man who said cats were more easily done than other animals, because they were so 'electrical'!

While some looked to mesmerism, others sought an alternative explanation of life's origin, its huge diversity and its way of working. The place of humans in the order of nature was the subject of philosophy and art, music and literature, and some members of these separate worlds considered the ordering of nature and its consequences more seriously and creatively than leading theologians. On mainland Europe art and science were also in conflict. Friedrich Nietzsche's first book, *Birth of Tragedy*, was published in 1872 and Richard Wagner was completing the *Ring* cycle for its first performance. They both wanted to combine beauty and the sublime, to promote the notion of raising human beings to an ecstatic level of being, celebrating the superman.

Darwin wrote on the subject to his American friend, the botanist Asa Gray:

> Your question [of] what would convince me of design is a poser. If I saw an angel come down to teach us good, and I was convinced from others seeing him that I was not mad, I should believe in design. If I could be convinced thoroughly that life and mind was in an unknown way a function of other imponderable force, I should be convinced. If man was made of brass or iron, and no way connected with any other organism which had ever lived, I should perhaps be convinced. But this is childish writing. I have lately been corresponding with Lyell, who, I think, adopts your idea of a stream of variation having been led or designed. I have asked him whether he believes that the shape of my nose was designed. If he does I have nothing more to say. If not, seeing what Fanciers have done by selecting individual differences in the nasal bones of pigeons, I must think that it is illogical to suppose that the variations, which natural selection preserves for the good of any being, have been designed.

Darwin stuck firmly to science and believed that heredity came from physical forces or chemicals discretely fixed into an organism's body, most likely as some small particles.

Experiments under glass

With the publication of the *Origin,* Darwin wanted to get on with his life. In 1859 he had turned fifty and he had so much more to do. To provide proof for his ideas of natural selection was the grand prize and he had designs for a great range of different experiments to test for the evidence he needed. Science and technology were moving ahead at a great pace, electricity especially revolutionizing industrial and domestic life. So far, the changes were based on physics and chemistry but some scientists expected it soon to be biology's turn. Darwin was excited about all the new toys being invented and he bought a big Beck microscope, machines to make thin sections and gauges to measure changing temperatures, weights and sizes: 'I cannot fancy anything more perfect than the many curious contrivances.' The equipment may have been up to date and the experimenter was wise and confident, but the work could only make a superficial mark on so many very deep processes.

Nevertheless, the new technology impressed Emma and Etty who were especially involved in operating the scientist's new Wardian case. This contraption was made of glass and metal and stood in the bay window of the dining-room. Inside, the chamber was kept warm and humid by tanks of boiling water in the base and from sunlight through the sides. It was to allow warm-temperate plants to stay alive through the winter and Etty first planted begonias and a strange variety of *Oxalis*. But the plants died at the first hard winter frost outside and the Wardian Case was given up as a failure. This meant that Darwin had only temperate plants for his experiments and that they were restricted to the seasonal patterns of growth.

He was determined to improve on this and to have better control on the many climatic variables that influenced his experiments. He found the money to add a heated extension to the already overcrowded greenhouse. It was to be a state-of-the-art structure with a coke stove and a double row of 6-inch (15-cm) pipes for the hot water running the length of the exposed glass side facing south. 'I long to stock it, just like a school-boy' he wrote to Hooker when it was finished in February 1863. 'Would it do to send my box-cart early in the morning, on a day that was not frosty, lining the cart with mats and arriving here before night? There would be about five hours on the journey home.' Darwin's driver brought 160 plants to Down House from Kew.

Asa Gray also sent him seeds of wild cucumber from the botanical garden at Harvard, a native plant of north-eastern America used to treat diabetes. Darwin knew that their tendrils coiled to the touch and hoped that close observation would find out how they climbed and grew round other stems. When he put a stick beside a branch the tendril touched it and then the branch began to respond separately. The tips of the branches searched like a radar scanner while the tendrils touched and curled around their object. Perhaps this would hold important clues to the processes of how a plant grew. When the seedlings started to germinate, Darwin saw that the 'uppermost part of each branch . . . is constantly and slowly twisting round'. The branch moved in a circular direction, describing a full circle every couple of hours, and then 'it untwists and twists in the opposite direction'.

On the Movements and Habits of Climbing Plants was published in 1865. Here he described the cucumber: 'The spontaneous movement of the tendrils is independent of the movement of the upper internodes, but both work harmoniously together in sweeping a circle for the tendrils to grasp a stick.' The work had taken him into the contemporary literature, mainly from German plant physiologists: 'These books stirred me up, and made me wish for plants specified in them.' Then he had passed problems for further investigation over to the experts at Kew Gardens.

In 1864 Darwin wrote to Hooker: 'The hot-house is such an amusement to me, and my amusement I owe to you, as my delight is to look at the many odd leaves and plants from Kew.' He wanted to cultivate the sundews that he and Emma enjoyed collecting from Ashdown Forest, south of Downe. They had first gone there in 1860 looking for bog orchids but instead found other plants that needed insects for food rather than pollination. Sundews, *Drosera capensis*, were carnivorous herbs with fleshy hairy leaves about half an inch (a centimetre) in diameter. Darwin suspected that the sundew had developed its sophisticated mechanism for trapping the insects in order to obtain their nitrogen which was absent from the acid soils of the forest's swamps. The plants' leaves were triggered to trap flies and spiders, and had rows of hairs secreting solvents to break down the prey into nitrogenous chemicals that were absorbed into the sundew's structure. Darwin was fascinated by this sophisticated system and thought its elucidation might reveal clues about the causes of other biological processes.

He didn't know what kind of agent he was searching for, but he was confident that somewhere within complex living processes such as these there would be clues about what made life work. For instance, were the processes involved with physical and chemical sensitivity in plants such as cucumber and sundew likely to lead to the discovery of units of inheritance? The possibilities became endless and Darwin began to investigate the sundew's touching mechanisms with a strange range of stimulants such as raw meat, camel-hair brushes and chloroform. In the house, Emma was surprised at the attention: 'At present he is treating *Drosera* just like a living creature, and I suppose he hopes to end in proving it to be an animal.'

Darwin's investigations went down a number of blind alleys:

> The point that interests me most is tracing the nerves! Which follow the vascular bundles. By a prick with a sharp lancet at a certain point, I can paralyse one-half the leaf, so that a stimulus to the other half causes no movement. It is just like dividing the spinal marrow of a frog . . .

But his work was precise and certainly his enthusiasm was unabated:

> I am frightened and astounded by my results. I declare it is a certain fact, that one organ is sensitive to touch, that in a weight one thousandth of a grain . . . suffices to cause a conspicuous movement.

The work attracted a lot of serious scientific attention and was reported at the 1873 meeting of the British Association for the Advancement of Science. Some of the plant's leaves had been macerated and their chemical composition analysed, opening the way to the first understandings of biochemical pathways. Hooker was particularly excited for it meant that plants were winning the levels of attention normally reserved for animals: 'Not merely then are the phenomena of digestion in this wonderful plant like those of animals, but the [ways they move] agree with those of animals also.' In 1875 Darwin's *Insectivorous Plants* was published filled with detail about the ecological and anatomical devices of these peculiar plants but with no answers to the central causes of the process or the enigmatic agents that Darwin was seeking.

Soon afterwards his son Francis, by then involved with his father's work, hinted at 'some matter' that was sensitive to light in part of the growing stem and which he said stimulated activity to bend it to the light. By then they had started to use grass and oats as experimental plants because they were easier to grow and their stem movements easier to monitor. The chemicals were isolated by others much later in the twentieth century, and were found to comprise a great family of auxins and other substances. But it had been another false trail in the search for the causes of biodiversity.

What these experiments couldn't test was the speed of evolution. To find out whether changes happened gradually or catastrophically Darwin needed to find out much more about the agents of inheritance and how quickly they could work. There were also the questions of how old the planet actually was, how long life has been happening on it, and whether the rate of evolutionary change was steady through all that time. The more Darwin thought about Lyell's insistence that change was gradual the more he realized that the concept was too simple to explain everything going on in nature. Nonetheless it was going to be difficult to prove because it was becoming clear that so much was going on at once in the whole complex system of life on earth.

Experiments with rabbits

The Tree of Life was a strong metaphor to describe the shape of evolution and Darwin's experiments were devised to test the theories that supported it. To find the driving force for adaptation and forward branching, he had to find the transmission agents. With his cousin Francis Galton he was searching for new ways of looking at how one species might suddenly flip and become another.

Francis Galton and Charles Darwin had the same grandfather, the great Erasmus Darwin, but had little contact until the 1860s. Their common descent was none too obvious and the two men turned out to have very different personalities, even though they learnt how to get on politely together. Where Darwin was slow and methodical, a gentle and kind, family man, Galton was fast and ebullient, apparently rather grim and without children. Charles was able to stay with a project for much longer, see it through to detailed enquiries and present comprehensive results and interpretation afterwards. But the cousins had the same broad view of science and an insatiable appetite for experimentation. They were both in their element when they were testing and measuring.

It was not until he was forty-seven years old that Galton published anything substantial. *Hereditary Genius* arrived in 1869 with the bold announcement that the book was 'the first attempt to investigate the special subject of the inheritance of human faculty in a statistical manner and to arrive at numerical results'. Partly encouraged to write the book by his visits to Down House, Galton's unstable mental health also had an influence: 'I often feel that the table-land of sanity on which most of us dwell, is small in area, with unfenced precipices on every side, over any one of which we may fall.' In this crude and naive style Galton had unwittingly started the pursuit of what was to become psychology, statistics and eugenics that determined the main political trends of the twentieth century.

At Down, Darwin was struggling with his own theory of the mechanism of heredity. He was desperately trying to discover how characters were inherited, the pangenesis that he had discussed in his 1868 work *The Variation of Animals and Plants Under Domestication*, where he proposed that an organism's structural characters were somehow transmitted from one generation to another. It was by what he called the gemmules, some kind of particle or agent that was collected in the sexual organs. In February 1868 he wrote to Hooker: 'I fully believe that each cell does actually throw off an atom or gemmule of its contents.' But all evidence eluded him, both of the particles themselves and any associated 'grand classes of physiological facts'.

This anticipation of the grand processes of the genetic code and protein synthesis that were to be worked out a century later was given without knowledge of a much earlier use of the same concept. He was told of this in a letter from a Dr Ogle who worked for the Registrar-General and who introduced Darwin to Aristotle's *On the Generation of Animals* and Hippocrates' *On Airs, Waters and Places*, dating from around 460–370 BC. Hippocrates wrote in the latter:

> If, then, children with bald heads are born to parents with bald heads; and children with blue eyes to parents who have blue eyes . . . What is to prevent it from happening that a child with a long head should be produced by a parent having a long head?

This like-beget-like phenomenon was explained entirely by what happens at conception. It was the simple version of pangenesis, when semen and eggs met, fused and formed a new organism. In March 1868 Darwin replied to Dr Ogle:

> I wish I had known of these views of Hippocrates before I had published, for they seem almost identical with mine – merely a change of terms – and an application of them to classes of facts necessarily unknown to the old philosopher. The whole case is a good illustration of how rarely anything is new.

Then the proud yet modest scientist in him went on:

> Hippocrates has taken the wind out of my sails, but I care very little about being forestalled. I advance the views merely as a provisional hypothesis, but with the secret expectation that sooner or later some such view will have to be admitted. . . . I do not expect the reviewers will be so learned as you: otherwise, no doubt, I shall be accused of wilfully stealing Pangenesis from Hippocrates, – for this is the spirit some reviewers delight to show.

Darwin and Galton were hoping to take the knowledge further with their work together at London Zoo. Galton was given facilities there, just across the park from where he lived. He worked out with Darwin how to give both male and female silver-grey rabbits blood transfusions from another breed of black rabbits. They then crossed these males and females to find what other characters their offspring inherited. In December 1869 they were breeding 'a few couples of rabbits of marked and assured breeds'. If gemmules really were in the blood as Darwin thought, then some features from the black rabbit donors should show up in the subsequent offspring.

Eventually there was 'good rabbit news' from Galton who had at first been confused by the transmission of some distantly related features. But when the breeding had covered several more generations and they found that of eighty-eight rabbits in thirteen litters none had any alterations in their breed, they became dispirited. In March 1871 Galton declared at the end of a lecture about the project to the Royal Society, 'the doctrine of Pangenesis, pure and simple, as I have interpreted it, is incorrect.'

When Darwin heard reports of the lecture he was furious; feeling let-down and even abused he wrote to *Nature* saying so:

> When, therefore, Galton concludes from the fact that rabbits of one variety, with a large proportion of the blood of another variety in their veins, do not produce mongrelised offspring, that the hypothesis of pangenesis is false, it seems to me that his conclusion is a little hasty.

A heated debate followed in the pages of this newly founded magazine but the search for particles in the blood didn't help find units of inheritance. Ten years later Galton was still talking of strange things like 'germs, residues and stirps' (a stirp is a family branch).

Emma was patently amused by the two men working together for she thought of Galton as a close family member and not a remote scientist. She wrote about the project to her daughter Etty:

> F. Galton's experiments about rabbits (viz injecting black rabbit's blood into grey and vice versa) are failing, which is a dreadful disappointment to them both. F. Galton said he was quite sick with anxiety till the rabbits' accouchements were over, and now one naughty creature ate up her infants and the other has perfectly commonplace ones. Charles wishes this to be kept quite secret, as he means to go on, and he thinks he shall be so laughed at, so don't mention [it].

It was, however, becoming very difficult to generate interest in these new ideas of evolutionary changes by the three features of natural selection, adaptation and divergence. What was wanted was a debate about how they interacted when the environment changed gradually and suddenly. More likely, most changes may have happened somewhere in between these extremes, whether they took hundreds of millions of years or the time of a single cell dividing. Perhaps it was inevitable that, without a new generation of evolutionary biologists, there should be negative reactions to such bold theories. But the spirit of the times didn't favour biology. Instead, most attention went to the popular technical innovations that swept High Victorian society. As well as Brunel's magnificent engineering there were to be the more domestic benefits of things like electric lighting and motor cars.

Physics' first threat to biology

Darwin was not the only scientist pushing the boundaries of knowledge. William Thomson, later Lord Kelvin, was a physicist intent on measuring the scale of geological time. The 'odious spectre' of Lord Kelvin had haunted Down House from 1866 when Kelvin had published a one-paragraph paper, with an appended calculation, entitled 'The "Doctrine of Uniformity" in Geology Briefly Refuted'. The phrase 'Doctrine of Uniformity' referred to Lyell's major inspiration in geology, and for Darwin the article's conclusion that the planet earth was only a hundred million years old, added insult to injury. Kelvin had measured the interior temperature and had worked out how the earth's crust restricted cooling from the molten state. He went on to refine this first estimate of its age to about twenty million years. Clearly, there was no time with Lyell's uniform current rates and gradual processes for evolution by natural selection to occur.

Darwin alone was adamant that Kelvin's data and assumptions were wrong and he was even sceptical of the calculations. They were much shorter times than his own estimates based on how long it had taken for the Weald of Kent and Sussex to have eroded. Wallace and Huxley agreed with Kelvin's dates, granting that some geological processes take place faster than others, some even catastrophically. In his final years even Charles Lyell himself had conceded that he might have been wrong to suggest that the present was a *universal* key to the past. But Darwin was concerned that Kelvin had insufficient evidence to justify the short dates. In his new calculations there wasn't enough time for life to have evolved, especially in its early stages. The transformation from the first living molecules to the first trilobite needed more than a few tens of millions of years. It was a simple clash between the evidence from physics and the evidence from biology and Darwin adamantly put all his money on biology.

It was not until the twentieth century that Darwin's ghost was laid to rest. During a lecture given in 1904 Lord Rutherford announced a much older age of the earth than anything advocated by Kelvin. After the lecture, Rutherford described his shocked reaction to seeing Lord Kelvin in the audience as he began the lecture:

> To my relief Kelvin fell fast asleep, but as I came to the important part, I saw the old bird sit up, open an eye and cock a baleful glance at me. Then a sudden inspiration came, and I said Lord Kelvin had limited the age of the earth, provided no new source of heat was discovered. That prophetic utterance refers to what we are now considering tonight, radium!

Just as Kelvin saw through Lyell's shaky edifice of uniformity, Rutherford had seen through Kelvin's single-tracked thinking about the cooling planet earth.

Even before this revision of the age of the earth to something which would have been much more acceptable, Darwin relaxed his attitude to Kelvin's fixed views and simply begged to differ. A few years before he died, Darwin's astronomer son George published an important paper about the age of the earth, 'On the Influence of Geological Changes on the Earth's Axis of Rotation'. News reached the author's father that even Kelvin was impressed, though the article was about a factor that he had completely neglected from his calculations. Darwin wrote to George in 1878:

> How this will please the geologists and evolutionists. That does sound awkward about the heat being bottled up in the middle of the earth. What a lot of swells you have been meeting and it must have been very interesting. Hurrah for the bowels of the earth and their viscosity and for the moon and for the Heavenly bodies and for my son George (F. R. S. very soon).

The garden's dark secrets and Francis Galton's rolling sphere

The still-hidden secrets in the garden at Down were now entering the next phase of history and the unhappy manner of their detection was to lead to a new kind of misunderstanding about evolution. Galton was an important spokesman for the X Club and he saw the whole of nature working together so that adaptation and natural selection followed automatically. As religion had diverted attention from this complex web during the nineteenth century so some of Galton's ideas would do in the twentieth. Together with the new pressure for scientists to measure and analyse there was a new trend for slow political reform.

Galton's book *Hereditary Genius* was an extreme interpretation of the inheritance of intelligence based on his biased observations about the intelligence of people in society. Underlying many of the interests that he discussed there was his simple statistical analysis of data to give a summary of variation across its whole range. It is the same kind of pattern shown in my 2002 book *Extinction: Evolution and the End of Man* for the changing diversity of species through geological time. It was taken up again in James Surowiecki's *The Wisdom of Crowds* in 2006. Galton had found the same pattern when he summarized data from many other different sources. The familiar shape of the bell-curve gives shape to the image of how the strong rise to a peak followed by a slow fall to extinction or some other oblivion. Whether using data from the topography of the Namibian landscape, meteorological readings with their changes of temperature and rainfall, or studies of the relationship between craniology and intelligence, the bell-curve gives meaning to the rise and fall of a number of different phenomena.

A year after his book was published Galton summed up his values in an address to the new Sociological Society:

> What nature does blindly, slowly and ruthlessly, man may do providently, quickly and kindly. As it lies within his power, so it becomes his duty to work in that direction. The improvement of our stock seems to be one of the highest objects that we can reasonably attempt.

But it was all to go horribly wrong as the history of the twentieth century showed.

Another controversy came a couple of years later with Galton's publication of a set of booklets called *Statistical Inquiries into the Efficacy of Prayer*. They were produced in response to the Day of Intercession, a national holiday controversially set aside by the Tory government to pray for Prince Albert's health. Galton's articles stimulated an angry national debate while his arrogance and simple disregard for the feelings of many people fuelled the fire. His argument went something like this: the average life expectancy for doctors was 67.07 years, lawyers 66.51, the clergy 66.42. 'Hence the prayers of the clergy for protection against the perils and dangers of the night, for protection during the day, and for recovery from sickness, appear to be futile in result.' Furthermore, he argued, why don't insurance companies distinguish between the pious and the profane on their application forms? Why don't doctors prescribe prayer for patients? Why do churches have lightening conductors on their steeples? A critical review of the pamphlets appeared in *The Spectator* and that in turn filled the office of the magazine with hundreds of angry letters. Darwin wrote to Galton about 'the tremendous stir-up your excellent article on "Prayer" has made in England and America.'

In 1902 the Royal Society awarded Galton the Darwin Medal and the President described *Hereditary Genius* as

> the first attempt to investigate the special subject of the inheritance of human faculty in a statistical manner and to arrive at numerical results – but in it exact methods were for the first time applied to the general problem of heredity on a comprehensive scale. It may safely be declared that no one living had contributed more definitely to the progress of evolutionary study, whether by actual discovery or by the fruitful direction of thought, than Mr Galton.

One of the ideas at the end of Francis Galton's 1869 book has also elicited a positive response from modern evolutionary biologists; some, like the essayist and palaeontologist Stephen Jay Gould, have seen it as a stroke of genius. It was typical of Francis Galton who was usually arrogant and naive about human nature, occasionally foolish and sometimes brilliant. Galton differed from Darwin with the view that evolution was not steady and gradual but subject to sudden bursts of activity.

Gould called this particular idea 'Galton's polyhedron', a metaphor of a rough sphere rolling in jerks, flipping from one stable position to another. The model represents intervals of stasis in evolving species or their very gradual change. The theory went much further than Darwin's more cautious hypothesizing ever dared.

A question being asked even in Darwin's day was whether Galton's gentle flip could ever gather momentum and assume catastrophic scales of change? In 2008, the metaphor is still useful.

The different answers have given rise to many schools of thought about evolution, pure Darwinists and neo-Darwinists to name but two. Thomas Kuhn, the philosopher, and Steve Gould were among the students at Harvard who argued that change happened between intervals of stasis, through shifts from one state to another, some stimulated by big catastrophic environmental events. The meteorite strike at the end of the Cretaceous period 65 million years ago and the earlier severe global volcanicity at the Permian/Triassic boundary both caused serious evolutionary crises. Such shifts are rare and disrupt the normally static history of nature and of science.

Together with Niles Eldredge, Gould saw evidence for such sudden shifts in change in the species of North American trilobites. More recently, Peter Sheldon, from the University of Wales, had different data and different interpretations. The two begged to differ about the meaning of 'sudden' but the definition of 'change' remained the same for both.

Most evolutionary biologists now agree that the history of the world comprises quiet phases, separated by catastrophic flips, 'evolution by jerks' as some cynics still prefer to call Gould and Eldredge's theory of punctuated equilibrium. This happens when Galton's sphere flips over one of its rough edges, moving to another flat surface, another quiet time of little or no evolutionary changes.

But were Galton's moments of catastrophe important causes of some evolutionary changes? While Darwin had laid down his theory of natural selection, and the pace of evolutionary change would continue to be debated well into the twentieth century and beyond, the seeming quiet of the garden at Down House had still to reveal the mechanism by which inherited characters were transmitted. While the search would continue to occupy Darwin until his death in 1882, its resolution would also become one of the greatest scientific discoveries of the twentieth century. Beyond into the twenty-first century there also appear to be links between sudden environmental change and extinctions of species.

Part Two

9

The Holly and the Ivy

Changing continents

In the bushes beside the Kitchen Garden and all along by the Sandwalk there was a surprising integration of pagan tradition and Christian sentiment that Darwin himself planted: the holly and the ivy are primitive symbols of male and female fertility, and used in some religious ceremonies to show the paradox of being. In the Christmas carol named after them, a line describing 'The rising of the sun' gives the image of an ancient fertility dance well rooted in some human paradise. But the master of Down House had a more functional use for these plants and they still use them in the garden borders to keep off intruders and to bind the hedgerows together. These climbing plants kept things in their place, set an order to the garden and protected what was going on there.

Recently it has been discovered that both holly and ivy have their deep history in the flora of the Canary Islands, the first port of call on Darwin's *Beagle* voyage. The holly enjoys the oceanic climate and one species of ivy grows there much bigger than anywhere else. It has also been declared that most of the 400 and more living species of holly now grow only in South America. In all these species, biochemists have found that the most common sequences of DNA are those found together in the Canary Island species and so that is presumed to be the oldest. But as with so many problems in biology, the more one finds out the more questions are raised, and then the more confusing the overall picture becomes.

After publication of the *Origin* Darwin became more certain that the things he had argued were right; in particular that migration could lead to the creation of new species. The discovery that had led him to shout 'Eureka' from his carriage in the lane up to Downe would have a lasting impact on all his ideas. The holly and the ivy would be a further part of his insights.

By the time he came home from his world tour in 1836 Darwin had been sure that the number of species on each island was dependent on the size of the region – the smaller the island the smaller the number of species – and that all ecosystems reach an equilibrium between the two factors. With such uneventful stasis it is no wonder that Darwin was struggling in the dark to find clues that might explain the dramatic shifts necessary for intercontinental migration. He was suspicious that catastrophic events such as those he saw in Chile might change environments but there was no known link between such earthquakes and volcanoes and geographical change. That major breakthrough was not to come for another century, when geophysicists could prove that continents drifted across the planet as tectonic plates. Meanwhile Darwin's contemporaries could only speculate about whether and how species migrated between islands and continents.

So to try and discover trends in what would later be termed biogeography, in the late 1840s Darwin once again busied himself with cottage-industry experiments, soaking fruits and seeds in sea water and opening the gizzards of migrating birds to identify where they last ate. He spent a lot of time and energy with experiments on seed dispersal from many species, fruits of different size and texture. It was clear to most onlookers that he was getting nowhere with this approach and needed to take more risks. Little did he realize that two important clues were much closer to him than these outside chances: distribution maps of the holly and ivy plants that were growing so flourishingly outside his greenhouse. They were to be among the first to show good evidence of their migratory history.

While Darwin was experimenting in his garden there was a younger charismatic naturalist more prepared to take risks about the ways that animals and plants had migrated from their point of origin across the continental barriers. Edward Forbes, the son of a wealthy family on the Isle of Man, was extrovert and speculative. He had joined several far-away expeditions, looking for plants and marine invertebrates from the Aegean islands. The way Forbes presented his ideas and himself was the opposite of Darwin and what he had to say seemed to be in conflict as well.

In 1840, when he was twenty-five years old, Forbes was elected professor of botany at King's College in the Strand. He then moved to the Museum of Practical Geology in London (later the Geological Museum) and finally to Edinburgh where he died aged thirty-nine. He was a bon viveur and attended some of the Darwins' early weekend house parties at Down. Forbes had recognized several European and African plant species on the Canary Islands and he was thinking of reasons why they were connected. One of his explanations was that they were from a sunken supercontinent that had once linked mainland Europe to the north of Ireland and south and west of Portugal towards the Azores. It was a bold declaration in contrast to what Darwin was doing, and the older man continued to be adamant not to approve of anything new without evidence.

Forbes was not only a good communicator but also he had a shrewd eye for structural detail. The Duke of Argyll, one of Darwin's critics, owned most of Mull and was interested in the fossil leaves being exposed on a cliff edge there at Ardtun. He sent specimens to Forbes for identification and was surprised and excited to hear they were unknown species of plane and oak. His theory was confirmed in 1988, when Zlatko Kvacek and I compared all the extinct species we found there to others from Ireland, the Faroes, Svalbard and Greenland and produced the first firm evidence that the same vegetation had covered the land between these places so long ago. The evidence proved that plants had migrated over the land connection between mainland Europe and America 55 million years ago. As the northern Atlantic Ocean started to open up, the islands such as the Faroes retained fossil evidence of the different environmental and evolutionary changes that had taken place there.

Just before Christmas in 1845 Darwin gave a weekend party at Down and invited 'the four most rising naturalists' to talk about biogeography. There was Hooker, to represent botany, George Waterhouse, an architect who collected insects, Hugh Falconer, a palaeontologist at the British Museum, and Forbes. None of these guests believed in evolution by natural selection, Forbes especially refusing to accept any 'real or bodily change' of one species into another. He was content to accept that species were divine, incarnate and acceptable inhabitants of Plato's otherworldly Atlantis. But Darwin was concerned that what Forbes was saying about islands challenged his own scientific observations: Forbes was far too romantic to be taken seriously.

It was not only that, but the religious determination Darwin's guests discussed as they walked in the garden at Down House frustrated their host in his hardline scientific quest, different as could be to Darwin's view that selection took place naturally. Throughout the weekend Hooker held 'aloof from all speculation on the origin of species'; instead he held 'the old assumption that each species has one origin and is immutable'. Hooker also argued that the sea transport of seeds to disperse plants to and from islands was inadequate and 'had been ridden to death'. The four guests were laughing at Darwin's experiments that soaked seeds in sea water to find how many days they could survive afloat and still germinate.

It was the beginning of a new tradition in biology in which hardline scientists kept strictly to the methodology of experiment and analysis of data around one variable at a time while the holistic thinkers considered more than one factor at once and so they were forced to speculate, especially when the techniques available were limited. Forbes was such a person, with a humorous and mischievous imagination that broke through the rationalism of the majority of contemporary scientists, anxious to take risks and make big claims. In his theory, a land bridge over which species migrated extended from western Africa out over the Canaries and the Atlantic Ocean to America.

Forbes continued to argue that plants go from 'one centre of creation to another' more easily by land than by water. Darwin was sceptical but he was a lone voice. His anger at Forbes's success with the idea of land bridges became too much in June 1856 and exploded in a letter that he wrote to his geology mentor Sir Charles Lyell. On the surface, Darwin was angry with the way that Forbes was using the myth of Atlantis to collect island biogeography as support for the religious ideas of animals and plants surviving catastrophic floods in the Ark. It was a fundamental reason why he favoured Lyell's gradualism so much. With the benefit of history, now it looks as though it was also to do with envy. At heart, Darwin was afraid that Forbes was getting too much attention, that he was a lesser man with unproven ideas and, what was more, he was winning a lot of the arguments. It was against such a background that Darwin eventually lost his temper in his letter to Lyell:

> I am going to do the most impudent thing in the world. But my blood gets hot with passion and turns cold alternately at the geological strides, which many of your disciples are taking.
>
> Here, poor Forbes made a continent extending to North America and another to the Gulf. Hooker makes one from New Zealand to South America and around the World to Kerguelen Land . . . And all this within the existence of recent species! If you do not stop this, if there be a lower region for the punishment of geologists, I believe, my great master, you will go there. Why, your disciples in a slow and creeping manner beat all the old Catastrophists who ever lived. You will live to be the great chief of the Catastrophists.
>
> There, I have done myself a great deal of good, and have exploded my passion.
>
> So my master, forgive me, and believe me, ever yours, CD
>
> P.S. – Don't answer this, I did it to ease myself.

In those days, with hardly any evidence of drifting continents, let alone of how tectonic plates cover the earth's surface, it needed the lateral thinking of a character like Forbes to offer possible explanations of the global distribution of modern species. The modern understanding of how continents have moved and the sea floor has spread gives a much clearer idea of how related species migrated. Islands were part of these changes and served as stepping-stones on several routes of passage, not least the Canaries for holly between the Americas and Europe. Volcanic activity from the tectonic plates of the spreading sea floor began the formation of the Canary Islands about 10 million years ago. It means that most life on the islands has migrated from nearby west Africa during relatively recent geological time.

Nevertheless, 10 million years was long enough for some variation in the species to have evolved. Because each island offered slightly different environments and opportunities, several of the species are unique to each island. Although the basis of Forbes's reasoning was without any real foundation, his conclusions were correct. Most of the islands that are presently close to continents were recently connected, or else they have formed independently from volcanic activities under the moving sea floor.

More than 100 million years ago South America was part of the now-broken southern continent of Gondwanaland that also comprised Australasia, Africa, India and Antarctica. There is much new knowledge of sudden global catastrophe, mass extinction events and the like, mostly unknown until the late twentieth century. Though different, each event somehow disturbs the equilibrium of biodiversity and goes on to cause changes in the evolutionary processes such as selection that favours some mutants and other extremes at the edge of the island populations. Piecing together this spherical jigsaw puzzle, with its changing climates, migrating and evolving species, and unpredictable catastrophes from within and without, can only be done when it is seen as a whole.

A question of chance

If Edwards Forbes was the nineteenth-century protester, then Stephen Jay Gould played that same defensive role in the twentieth century, but with a different style. Both these scientists thrived on the speculations and jests derived from buried remains of life. Just as Yorick's skull reminded Hamlet of his boyhood, so buried continents were a useful theory for Forbes to make his case, and static fossil assemblages supported Gould's ideas of long geological intervals without evolution. It was what he called 'punctuated equilibrium'.

Stephen Jay Gould was a palaeontologist and essayist from New York who found the narrow rationalism of the English way of thinking particularly restrictive. He could see no reason why adaptation should be the only way that evolution worked, for sometimes chance also played a part, and there was the necessity of rejecting less fit individuals. These differences were highlighted in an unexpected encounter at the Royal Society in 1978, then recently moved to a new building in Carlton House Terrace just across the road from Darwin's club, The Athenaeum. The convenor was John Maynard Smith and he had invited Gould to talk about his latest theoretical work on the way organisms adapt to the environment.

Maynard Smith was about as hardline an English rationalist as you could get and enjoyed calculating the chances of victory in theoretical competitions he set for biologists. Rather than talk about the stable state in evolution, Gould brought a completely new explanation of how some structural features in his specialist group, the molluscs, might have adapted. He compared some of the shells' features with an unusual feature you find in churches, the triangular architectural void between two arches and the ceiling. The triangular space, or spandrel, was the inevitable by-product of these necessary supports and was adopted by artists as a convenient place for their paintings and sculptures. Gould wondered that day whether some structures in snails were similar by-products that go on to find some purpose by chance.

The analogy didn't go down well in this new cathedral of British science, still proud of Darwin's fellowship and his belief that adaptation is the central evolutionary mechanism. When we can measure adaptive changes in DNA and describe precise change at the molecular level, some trumped-up sideshow from architecture was not what Maynard Smith had in mind at all.

To most of the scientists in that audience chance was as unsatisfactory an explanation for why and how we are here.

I was there that day, listening to these arguments with the same surprised amazement as others. From the back I couldn't see the detail because Gould bent down in front of the stage, but I did hear him shouting at the chairman, the society's mollusc specialist Arthur Cain. Gould was telling him to look at the front of the lectern and read the society's motto, written there under its logo. Although it was done half-jokingly there was enough anger to make it very serious and a spell of insecurity fell across the room. Gould was in full flow, quoting the motto on the lectern *nullius in verba* and translating it as 'on the words of no one' to mean that scientific validity is established by experiment rather than accepted authority or received wisdom. Gould argued that his theories of biological evolution were doing just that. The English would do better to relax and open up to other possibilities than retain their obsession with adaptation alone.

Gould went much further in his 1,433-page unedited treatise on the history of thinking about evolutionary biology that was published just before he died in 2002. There he suggested that Darwin was confused about whether evolution took place when the environment was stable and whether extinction only took place after some kind of environmental catastrophe. Gould thought that Darwin had backtracked in the *Origin* from his earlier statements in 'The Big Species Book' that environmental change was a good thing for evolution. In 'The Big Species Book', back in the late 1850s, Darwin had implied that separation on the edge of communities or on islands encouraged adaptations to thrive. But would the parent forms, still in their original location, become extinct? New adaptations would surely tip the equilibrium of any organized system. Any benevolence would fill up too little space with too many different forms. Surely, some would have to go? Would change be by chance or functional adaptation? Gould argued that the answers were not clear to Darwin and so he stopped asking such questions in the *Origin* itself.

The edited version of 'The Big Species Book', published only three years before this meeting at the Royal Society, and its unexpected revelations were at the front of Gould's mind. Darwin had proposed that groups of organisms live or die by competition. As on any field of battle, that must be between the individuals themselves, while the victors would then gather together to split off as a new group. There was less competition when the space was less crowded, such as in extreme environments; but in such places the chance occurrence of adaptation could lead to unexpected success. But why and how should the new species replace the old? Gould argued that Darwin got 'bogged down' when he raised this possibility in 'The Big Species Book' and said nothing of it in the *Origin* or anywhere later in public. It was as though he was unwilling or unable to accept the demise or extinction of the losers.

'The Big Species Book' has two diagrams showing the principle of divergence when any one species splits into two and is forced into competition. One shows the varieties increasing in number and diverging in character; the other was put aside by the time Darwin approached the *Origin*. Now one can appreciate that this diagram is the most interesting and certainly the most creative. It postulated that species at the edge of a population have more variety and more scope for divergence: extreme variants show

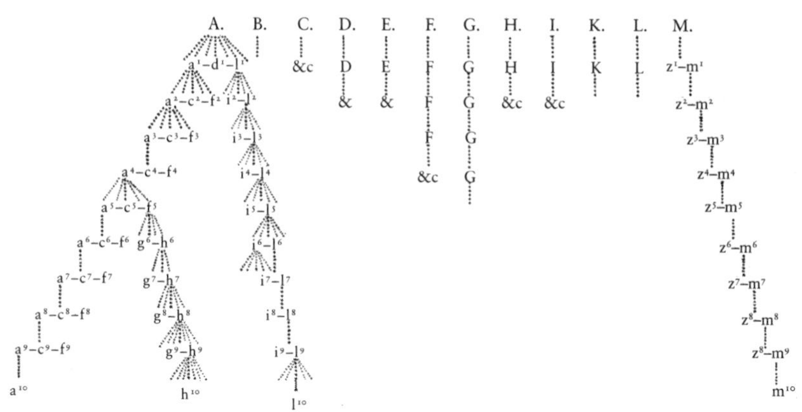

Charles Darwin's Natural Selection: Being the Second Part of his Big Species Book Written from 1856 to 1858, ed. R. C. Stauffer, Cambridge University Press, 1975

this trend while modest normal ones do not. The diagram was a more thoughtful description of Darwin's eureka realization, in his carriage up to Downe, that geographical separation can lead to a new species when the environment is different.

Recently, there has been a lot of speculation about why Darwin chose to leave this second figure out of the 1859 *Origin*. Was it because he was uncertain about whether a newly adapted form retains all the ancestor's features? What is the difference between adaptation and selection? Should the branch of the original species stop or continue? Should they become extinct or adapt, or even stay the same?

When Darwin was writing 'The Big Species Book' he recognized this difficulty and saw that the individuals at the edge of a community were able to diverge in their structure if a chance mutation formed a useful change. On a bigger scale this became migration. In 1857 the Swiss botanist Oswald Heer had explained migration of plant species between Europe and America by the theory that a land bridge had existed around the time of the last glaciation. It was the same theory as that proposed by Forbes; now Heer also argued that the Canary Islands and the Azores were all that remained.

Holly's lost origin

During the 1990s my own research group accumulated a database of nearly all the known findings of holly in the fossil state, whether as wood, leaves, fruits or seeds and particularly the very distinctive pollen. We had information about the geological age and location of thousands of specimens that allowed us to make a series of world maps showing the changing distribution over millions of years. The sequence of plots showed a crude history of how holly might have evolved with the changing shape and placing of the continents up to the present. Unfortunately the fossils didn't show the kind of detail to allow any clear identification of species and only part of the story was preserved.

One day I had a call from Jean-François Manen, a molecular biologist at the botanical garden in Geneva. He had another set of plots from living species of holly all over the world, based on their sequences of small fragments of DNA, and he wanted to fit them into the picture emerging from the fossils. It was an exciting opportunity to fit together everything that was known about holly evolution. I remember walking with Jean-François up Parliament Hill, just north of central London, confident that we were going to see the whole history spread out before us. We had all the data available from biochemistry, anatomy, geography and geology, and all the expertise and technical facilities to put them together. Looking down on the vast expanse of London we were hoping for insights on all the origins and expansions since species of holly began. Or so we thought.

One problem involved a mismatch of the results from sampling different parts of the cells: the chloroplast DNA gave different results to the nuclear DNA. Jean-François put this down to a mixing of the lineages or some other kinds of hybridization, accounting for a lot of noisy information, grey data that didn't seem to mean much. But some of the analysis showed up as three distinct and colourful branches or clades, one restricted to South and North America, one to South America and East Asia and the third to Eurasia.

Another problem appeared from Australia where there was good evidence for holly species today and right back to about 5 million years ago. Before then, and back to the first record of the group about 80 million years ago, there were only a few dubious records. Their authenticity was being challenged and there appeared to be no really reliable evidence for any holly species more than 10 million years old in the entire southern hemisphere. But it was always difficult to conclude that things were completely absent and such early plants may have become extinct with no fossil evidence. This must have been the reason for another difficulty, the great gap between the earliest holly species and its closest ancestors. None of the evidence we had from DNA, anatomy or fossils, seemed to connect.

Eventually, like an image emerging through a misty morning, we were able to put a few of the vague threads together. Holly species have been most abundant in the northern hemisphere and go back there to the great Arcto-Tertiary flora of temperate forest. A lot of the evidence from the work of Manen and myself went along with the idea that *Ilex canariensis* is the oldest living holly species and may be close to where holly began. That would account for the one very clear observation in all the data: the very different chloroplast DNA found in Asian species to that from the North American species. It also set a good date on the origin of the holly clade, back to when the Canary Islands themselves formed. The main guide to this date was the latest geophysical estimate that the oldest island in the group formed about 10 million years ago.

These pieces of evidence allowed us to reconstruct the way in which holly might have evolved, geographically, ecologically and structurally, through the last 10 million years. When the first Canary Island began to form, the early species migrated west across the then narrower North Atlantic to North America. The continental plate of southern America was just in the process of joining with the northern continent and so the genus migrated down south and exploded with many new species. Meanwhile, to the east of the Canaries, other species of holly had evolved and migrated into Asia. To the north-west another cascade had touched the English Weald, where the chalk escarpment of Darwin's garden was by then well eroded. The cycles of new species of these and other plants and animals moved onwards like our species' social history and the changes of our culture. The holly and the ivy turned again to a pagan celebration of science itself.

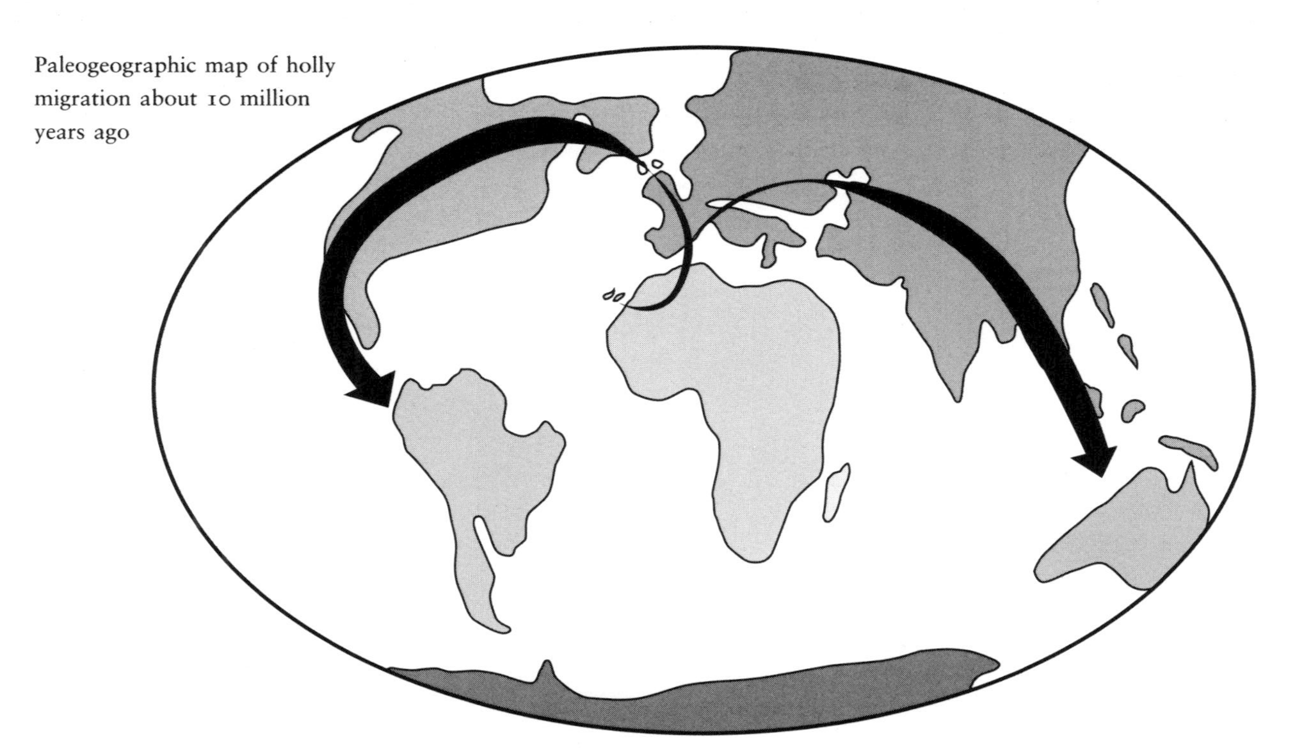

Paleogeographic map of holly migration about 10 million years ago

10

The Rise and Fall of Mendel's Genetics

Locked up in the garden

After all the fuss immediately after publication of the *Origin* and pleased with the warm response from so many admirers all over the world, Darwin was anxious to retreat into his garden at Down and take his ideas a stage further. He moved away from the pigeon-breeders at London Bridge and from the editors of *The Gardener's Chronicle* hoping for manuscripts from their well-known correspondent.

Throughout the 1860s his main concern was the search for evidence of the agents of inheritance, breeding the birds and flowers to find patterns in the ways they changed, not just to get fine specimens. He knew that they would be found by experimentation, not observation alone, and he was convinced that the answer would be found in something chemical within the cells. But he had no great confidence that he would find the elusive agents, as though he accepted that one great discovery in a lifetime was enough for one person. He even gave some of his friends the impression that he expected the clues to be found by someone else.

Darwin became obsessed with his experiments, looking for stable features in the new varieties of gooseberries and pigeons, controlling the pollination of primroses and orchids. In a letter to Huxley in 1865 about starfish he speculated about the likely hereditary units. Could newly evolved organs be formed from the 'free diffusion in the parent of the germs or gemmules'? He found it 'extremely difficult to reconcile on any other view' than from these 'quite remote spots'. The more free that diffusion was, the more out-breeding there was, the more successful the adaptations and selection became. This was the basis of many of the experiments that he conducted in his garden until the 1870s. Not for him the sudden flip of Galton's eccentric ideas or the other quick mutations that some of the young biologists were looking for.

The first two chapters from his manuscript of 'The Big Species Book' of 1858, in which he argued that the gemmule linked the body cells of one generation to the next, formed the basis of *The Variation of Animals and Plants Under Domestication* which was eventually published ten years later. The reviews however were bad. One of the leading literary magazines, *The Athenaeum*, said: 'Henceforth the rhetoricians will have a better illustration of anti-climax than the mountain which brought forth a mouse, . . . in the discoverer of the origin of species, who tried to explain the variation of pigeons!'

Darwin could find no evidence for the gemmule through the experiments in his garden, however. Instead he developed more sophisticated notions of pangenesis, about the way the still mysterious agent might be transmitted during sexual reproduction and the way the characters show up in development. Bravely he allowed himself to be led by his creativity and even suggested that the transmitted characters may not always be manifest in the new organism, that some would only show up at late stages in development or in future generations. He was convinced that pangenesis would eventually explain how such adaptation could be transmitted between generations as well as how changes such as mutations occurred.

In hope, in 1868, he wrote to Hooker:

> You will think me very self-sufficient, when I declare that I feel *sure* if pangenesis is now stillborn it will, thank God, at some future time reappear, begotten by some other father, and christened by some other name.

Darwin argued strongly that pangenesis was the process enabling evolution to happen even though he hadn't isolated its parts. Short of any evidence he resorted to jokes about the god Pan and J. M. Barrie's little boy Peter. He wrote to Huxley about the virtues of the concept, its encouraging pluralism and understanding that nature was always pressing ahead and would never grow up or reach any final form.

It was as though Darwin had got locked inside the garden with too many things to observe at once. The experiments with primroses held out so much potential, but one after another they resulted in nothing new. For a large and growing part of Darwin's motivation this didn't really matter because he found the projects so enjoyable. The complacency showed through in his correspondence to Hooker in 1867:

> The other day I had time to weigh the seeds, and by Jove the plants of primrose and cowslip with short pistils and large grained pollen are rather more fertile than those with long pistils, and small-grained pollen. I find that they require the action of insects to set them, and I never will believe that these differences are without some meaning.

He presented that work at a meeting of the Linnean Society the following year and afterwards wrote about how it made him feel to his friend Joseph Hooker:

> I by no means thought that I produced a 'tremendous effect' in the Linn. Soc., but by Jove the Linn. Soc. produced a tremendous effect on me, for I could not get out of bed till late next evening, so that I just crawled home. I fear I must give up trying to read any paper or speak; it is a horrid bore, I can do nothing like other people.

Since the publication of the *Origin* some of the new professional scientists who visited London also came to visit Darwin at Down House. In the summer the guests had tea out on the veranda, eating juicy fruit from the greenhouse and laughing with their convivial host about how he grew these foreign plants. But the main purpose of the glass laboratories was to germinate primroses and orchids before the hardy ones were planted outside and the visitors admired the range of different tests that were going on in such an ordinary place. As well as a peach tree, Darwin had grown a stray banana plant in the corner and admitted his own fun from breeding odd combinations of species together in the artificial surroundings.

Darwin was not alone in hunting down the units of inheritance that governed adaptation. Some kind of agent of inheritance had been contemplated before. The Greek physician Hippocrates had explained many illnesses presented throughout a family history as pangenesis, as Darwin had discovered from his correspondence with Dr Ogle. In France, a hundred years before Darwin was at Down, Georges-Louis Buffon had spoken of a physical 'mould' being transmitted from one generation to another. Yet these notions were of little value to Darwin's experimental approach. His search was for something inside the organism that needed to be uncovered.

He discussed the importance of the mysterious gemmule with many of his correspondents and visitors. Among his guests was Hugo De Vries, a bright young botanist from Amsterdam who, modifying Darwin's work, would name the units of inheritance pangenes, in his 1889 book *Intracellular Pangenesis*, somehow linking the inner parts of each cell. De Vries recalls meeting the master of Down towards the end of the great man's life in 1878:

> In the garden there were hothouses with peaches and grapes. Darwin told me a long story about the peaches and immediately offered me one of them; it was delicious. He has deepest eyes and in addition very protruding eyebrows, much more than one would say from his portrait. He is tall and thin and has thin hands, he walks slowly and uses a cane and has to

stop from time to time . . . His speech is very lively, merry and cordial, not too quick and very clear. It is remarkable how soon one feels at home with people who are friendly . . . It is such a pleasure to find that somebody is really interested in you and that he cares about what you have discovered.

Mendel's peas

In his garden at Down House the doubtful naturalist was still without evidence for the next stage and his work was not winning any further support. No experiment in the garden gave him the evidence he needed and he began to think that the proof would not be found in the garden. Nonetheless he stood his ground, convinced that adaptation by natural selection worked by a system of free diffusion. Although he made it plain many times that there could be other ways for evolution to happen, he was more and more convinced that this was all he could support with the information available to him. It would be another garden in the very centre of Europe that supplied the answer.

Unknown to Darwin and the small international scientific establishment, Gregor Mendel was breeding peas in his own, remote greenhouse. It was there in this far-away country of the Habsburg Empire that experiments were to show up the secret of genes. The work showed that some features such as colour and texture were transmitted to the next generation in rearranged combinations for each individual. They showed up in such a way that it was best explained by the existence of particles or gemmules inside the reproductive cells.

Mendel had been brought up in a farming family and was used to the conventions of animal and plant breeding so he wanted to learn more about the theories that were involved. This became possible after his ordination as a monk at the local monastery of St Thomas in Brno (now in the Czech Republic) when he was given the chance to study at the University of Vienna. He was twenty-five years old and full of enthusiasm to learn about the wonders of living things and became especially influenced by Franz Unger's lectures on plant breeding. But there were more theoretical and exciting new things to learn in Vienna from another botanist called Karl von Nägeli, who lectured on the phenomena of life in terms of physical and chemical laws. It was part of the growing belief in central Europe as well as in Britain that some kind of metamorphosis of species led to 'transmutation' or evolutionary change. Mendel worshipped his radical mentors whose ideas clashed with the established authorities and this may have been why he failed his examinations, by not saying or writing the right thing, forcing him to leave Vienna and return to the monastery.

At Brno he was well supported by the progressive Abbot Napp who gave him a big new greenhouse and what seems to have been great intellectual freedom. This was well appreciated by Mendel's friends and encouraged another monk there, Matous Klácel, to read the new German *Naturphilosophie*. This group, which had included the philosophers Friedrich Schelling and Georg Hegel, promoted the radical evolutionary thinking derived from Goethe, and propounded the idea that nature has a fundamental unity. There was much discussion of their ideas within the monastery walls between the curious men who gathered around Mendel. Primarily, they were anxious to find up-to-date answers to the questions from their fellow monks about the role of deep nature in evolution.

The popular science book of the day, the anonymously written *Vestiges* that had caused such a furore in Britain, soon became one of the most widely read and talked-about works in the monastery library and became the main subject of conversation among the professors in Vienna. In 1847, while at university, Klácel had got into trouble with the censor for attempting to publish statements such as 'every age has much that is transient, which the dialectic sifts and polishes, until its kernel is revealed'. The German-speaking intellectuals wanted to take Kant and Hegel further into line with their religious beliefs but the establishment didn't see these tests on nature as the right way to go.

Mendel's ground-breaking experiments in his own garden – counting the different kinds of offspring from crossing pea plants

– lasted from 1856 until 1863. The work involved crossing thousands of true-bred pea plants which bore either wrinkled or smooth seeds and counting the numbers of each character in the pods of offspring in the next generation, then repeating the exercise with a third generation. The ratios showed that the two characters were being mixed during sexual reproduction and Mendel proposed simple rules that he hoped would explain how this happened. His analysis was particularly clever because he assumed that each character had two states of coding, dominant and recessive. The dominant could hide the recessive as chemical code stored inside the cells of each plant. Somehow, the chemicals could control characters like wrinkled and smooth seeds. He lectured on the results in 1865 to the Natural History Society of Brno and then published them in the society's *Proceedings* in 1866. It was an obscure periodical and not many people outside the society read it.

Two years after the article's publication, Mendel was appointed abbot and stopped working with his peas altogether. His originality was not appreciated during his lifetime and although he died in 1884, two years after Darwin, Charles never knew of his work. It was only when De Vries wrote a draft manuscript in 1900 about Mendel's research and began to mention it to friends like the Cambridge don William Bateson, that Mendel was rediscovered and his importance recognized. This led to the birth of a whole discipline that flourished until the peak years of the 1960s when the genetic code was discovered by Francis Crick's group at Cambridge. The early genetical studies of genes recombining in pea characters such as wrinkled seeds led on to work on mutant genes expressed in freak flies and frogs.

Genes were to become the focus of twentieth century biology, the favoured unit of measurement in the new science of genetics. This began as the study of heredity by looking at morphological characters, then at the chromosomes, mutations, selfish genes, triplet codes and, most recently, what some call cell regulators. It was the way of finding more clues to the secrets of life than had been discovered before. Since then looking at the complex role of DNA within the matrix of biochemical pathways has taken over.

Legend has it that Bateson read De Vries's draft manuscript about Mendel on the train to London where he was to be guest speaker at the opening ceremony of the 1900 Chelsea Flower Show. The journey gave him enough time to read De Vries's article thoroughly and it made an enormous impression on him. The article set out a clear summary of Mendel's work and De Vries's conclusions, and was the first that Bateson had heard of Mendel and his work in Brno. It made him realize in that railway carriage, in yet another eureka moment for evolutionary biology, that Mendel's ratios of characters in successive generations of pea plants were robust enough to account for features continuing to be inherited. In the horse-drawn hackney carriage from Liverpool Street station to Chelsea, Bateson decided to change the conclusion of his lecture to show his appreciation of Mendel.

Bateson invented the word 'genetics' for these studies of inheritance and variation, eventually becoming the University of Cambridge's first professor in the subject. But he had a hard career partly because he was best known to stand against ideas rather than for them, natural selection being one of many he disdained due to the lack of clear evidence. Like many professional academic scientists then he was not an approachable man, but he was beginning to see evolutionary biology as more than a set of black and white issues. Through the first decade of the twentieth century it seemed to people like Bateson that an old divide was being bridged, driven by the forces of science and technology working on the traditions of myth and feeling. There had been two accepted ways in which most people expected to see the world. One was the metaphor of human experience, a chaotic flux that somehow works. The other was the factual realism of the here and now, the view based on true measurable observation, the growing stuff of science. Bateson's was the first generation of biological scientists to cover both these approaches as professional thinkers.

The growth of genetics

The new discipline involved a search for clues to sudden changes, centred on the things that actually transmitted the hereditary features. Mendel had provided a physical explanation for natural selection with the experimental evidence to support what Darwin had only imagined. By sharing and mixing these things, the 'genes', their actual physical characters would be selected in nature; genes were part of a link between organisms and their environment. Their data could be found in many cells of the body, in the tissues of mammals, the flowers of plants, and exchanged sexually through the reproductive cells.

With so many technological advances the new twentieth century was the right time for a scientific approach to be used to investigate life sciences systematically. The dominant method was established at the start of genetical studies, with the emphasis on experiment, measurement and statistical analysis of the results. The new study of genetics encouraged by Bateson started with frogs and fruit flies but swiftly moved to work with smaller organisms such as relatively simple bacteria that had shorter life cycles and were quick to grow. In time, as technology improved, geneticists looked even further into chromosomes and the quest for the answers to life merged into chemistry and physics. Through the 1920s and 1930s it was exciting to see chromosomes inside the cells that could be extracted and prepared for the microscope. And it was expected that soon researchers would be able to see the genes themselves.

The First World War was a catastrophe that increased the value given to physics and chemistry over biology and this stimulated those who fancied their chances in the new discipline of genetics to reflect that trend in their own work. It was the time when philosophers of the Vienna Circle of positive thinkers such as Philipp Frank tried to absorb scientific methods into the more artistic history of human society. The logical positivists saw science as a set of statements reducible to simpler statements. At the time it left physics and chemistry stronger and the biology of natural history weaker. In this materialist world it was inevitable that progress in developing Darwin's ideas on adaptation by natural selection began to recede. Bateson summed up the mood to the American Association for the Advancement of Science in 1922:

> Less and less was heard about evolution in genetical circles, and now the topic is dropped. When students of other sciences ask us what is now currently believed about the origin of species we have no clear answer to give. Faith has given place to agnosticism.

During the interwar period the international stage of genetics was dominated by two pioneers in that field, J. B. S. Haldane at Oxford and Ronald Fisher in New York. Haldane stated his case with characteristic immodesty in his 1932 book *The Causes of Evolution*: 'I can write of natural selection with authority because I am one of the three people who know most about its mathematical theory.' From their mathematics Haldane and Fisher found evidence that newly formed characters spread in a population provided they are present in enough individuals of the population to prevent their disappearance by random extinction. Haldane and Fisher also argued that evolution occurred mostly by mutations and that these only occur along certain lines, for they mostly lead to a loss of complexity rather than an increase. Darwin would not have been happy but no one seemed to care.

Yet the mathematical approach to genes did not satisfy everyone. As a professional scientist Bateson had a lot of sympathy for modernist objectives, but he was also attracted to the other side of the divide, the world of feelings and of myth. Although he hid inside the cloistered conservatism of Cambridge he didn't much like the even more arrogant aggression of heroes such as Haldane. For Bateson, and many others, there was another approach to life, away from facts and measurement. This was perhaps why he regularly carried a copy of Voltaire's *Candide* in his pocket, exposed so that people could see the title. There was more to the stuffy old man than genetics. Like Darwin a generation before him, he was afraid to make up his mind about many of these deep phenomena.

It seemed that Bateson's hopes for a softer genetics, which embraced the diversity of whole floras and faunas, was going against the grain. This metaphorical side to the value and way of life has been overshadowed since Voltaire's day by an over-emphasis on collecting single facts. Throughout the twentieth century geneticists have been seeking smaller and smaller explanations for their hereditary phenomena. Evolutionists have turned into molecular biologists and have become very successful in understanding and applying many of the central processes involved. But sometimes this encourages over-simplified talk of the cancer gene and the obesity gene rather than the miracle of the whole genome working as one.

The Modern Synthesis that didn't work

An attempt to bring the many sides of a more holistic approach together began in New York during the 1930s when a challenge was made to the dominance of the study of molecules in biology. At that time Theodosius Dobzhansky, a geneticist from the USSR, and the entomologist Ernst Mayr from Germany had become excited about the wider issues of evolutionary biology. They joined with the vertebrate palaeontologist George Gaylord Simpson and the zoologist grandson of Thomas Huxley, Julian, who was later to become first Director-General of UNESCO.

The group's main aim was to bring natural selection back into mainstream biology and to cooperate with other disciplines in the field that were becoming isolated and specialized. In his 1942 book *Evolution: The Modern Synthesis,* Julian Huxley wrote:

> The time is ripe for a rapid advance in our understanding of evolution. Genetics, developmental physiology, ecology, systematics, paleontology, cytology, mathematical analysis, have all provided new facts or tools of research: the need today is for a concerted attack and synthesis.

By then the four scientists had each written an important book generally agreeing on the form and scope of Darwin's ideas, setting them in the knowledge of Mendelian genetics and contemporary ecology. They agreed about the effects on populations from migration, their gradual change through geological time and on the importance of adaptation for the good of a lineage.

The English interpretation emphasized adaptation, or 'functional guidance' as Julian Huxley called it, while the three Americans were more sympathetic to other ways for creatures to evolve. Chance had a big part to play, the impact of accidents, and trends such as similar structures being used in different places at different times. It led to the recognition of rogue elements that defied Mendelian inheritance, just the kind of possibilities that Darwin had emphasized. But there was no hint then of Galton's polyhedron.

The four evolutionists named their agreement the Modern Synthesis and celebrated with a conference in Chicago in 1959, the centenary of the publication of the *Origin*. They represented the pluralist world of science, showing that adaptation and mutation were not solely based on biochemical programming within the gene. The environment, accident and adaptation could also develop changes in a species.

One of their ideas was that out of the wreckage of both world wars there should be a less aggressive approach to the science of nature. It would bring together the high objectivity and measurement of mathematical analysis and the less precise world that Bateson and Voltaire appreciated so much. It was a long shot and at the time it seemed that the modern synthesis had not worked.

The Second World War didn't stop the rise of mathematical biology. Modern Synthesis, it seemed, had failed to open new horizons and the search for the key of life once more focused on the internal chemistry of cells as each generation delved deeper and deeper into the material of things. This came to the fore at University College in the 1960s, around the corner from the Darwins' old London house on Gower Street. Here Francis Crick, who had originally studied physics, lectured about his recent discovery with the American crystallographer James Watson of the genetic code.

That same year the biochemist François Jacob from Paris taught that proteins could block or unblock the same bits of DNA that synthesized them, like an electric light's on–off switch. He lectured at the Galton Laboratory, which was still in the same building as the Slade School where Gwen Raverat, Darwin's granddaughter, learnt her woodcutting skills. Looking out on to the front quad with its lawns and ginkgo trees, it was still much as it had been in the days of Galton's eugenics that led to the mathematical game theory of J. B. S. Haldane and his student John Maynard Smith.

These men all took the hard line of genetics to molecular biology. Today there are so many specialists who have delved further and further into the nature of genetics that even the word 'gene' is difficult to understand because it has so many meanings, at so many scales and levels. In 2005 it took two days for twenty-five specialists to agree one definition: 'a locatable region of genomic sequence, corresponding to a unit of inheritance, which is associated with regulatory regions, transcribed regions and/or other functional sequence regions'. More simply, a gene is a protein coding of DNA with no clear beginning or end and with a major role for RNA. Many of these scientists, chemists specializing in proteins or nucleic acid biochemistry, have displaced the culture of Bateson's Mendelian genetics, a way of thinking now long gone.

It is all an even longer way from the entangled bank at Down, let alone the untold joys of Orchis Bank on the walk to Cudham. Genes had eluded Darwin as he hunted for the units of inheritance that would help explain his ideas. But he had also been on a quest for other key notions to evolution in his garden.

II

Orchids Become Hopeful Monsters

Orchids at Down House

Orchis Bank was Emma and Charles's favourite spot on the walk down the hill from Down to Cudham, where the locals boasted that no British county excelled Kent in the number of orchid species. In the spring of 1844 Darwin scribbled his feelings about the place in a notebook, showing his excitement about the 'most agreeable songs on all sides, from the larks that abounded there and the nightingales'. Green-winged orchids grew with cowslips in the fields and the young couple brought back pots of the wild plants to transplant in the greenhouse. One species changed shape according to the type of soil that was used and another had to be kept away from the rest. Neighbours added to the collection as did specialists from Kew and overseas. Darwin watched with pleasure as he saw that each species showed its own cunning innovations to attract a particular pollinator. Some fitted snugly into the shapes of the specially adapted petals, others shot out their pollen masses if the insect touched the right part of the flower. They were convincing examples of the role of diversity within the environment; the integration between insects, flowers, fungi and soil bacteria, interacting to gain a balance.

Darwin noticed that, like hedge parsley, some orchid species changed their form and colour in regional patches and some of these migrated from one place to another. Like the pin-eyed and thrum-eyed primroses, orchids developed different inbreeding strategies as their surroundings changed. Like some of the inherited characters that were cherished in pigeons and peas, several notable features in orchid flowers attracted a lot of excitement among orchid breeders, especially when it became clear that the new adaptations were permanent. Such changes happened frequently enough to make most observers question whether evolutionary change was as slow as Lyell had persuaded them to believe.

The orchids were a group of plants that had interested Darwin as a boy and again as a young man exploring the Galapágos Islands. He had begun by thinking that because they were so sensitive to the environment, they evolved quickly, and so they are particularly diverse. In the early 1840s the expert botanist Robert Brown, whom Emma found so dull, showed Darwin, during his many visits to the British Museum just before the Darwins moved to Down, that the variations in the orchid Family were more complicated. In those days botanists thought there were about 6,000 species of orchid though now estimates average around 20,000, with more species than nearly all other Families. They also have more elaborate mechanisms of flower structure than most other plants.

Finding out why the orchid Family was so diverse was one of the projects that Darwin wanted to pursue once the *Origin* was comfortably out of the way. It was to turn out to be as disappointing a search as that for the gemmule, and not until some years after his death was it clear how there became so many species or what was the mechanism for their adaptation.

Darwin would lie on the grass to see the bees enter the flower, push into the spur in search of nectar, the very sensitive transfer of the pollen masses, and the graceful withdrawal of the insect's proboscis when it left the pollen. Bees and wasps were most often involved, but he recognized that flies, butterflies and moths and humming birds were also part of orchid life cycles. The flowers were part of a whole system within the plant that gave a vast range of colour, shape and smell. From such intricate performances as he observed fertilization often took several months to complete. Eventually the orchid fruit would release as many as 4 million tiny seeds, to be dispersed by the wind.

Darwin's friend Hooker was convinced that the fancy orchid flower was specially adapted to fertilize itself and thought there was no pollinating agent such as wind, mammals or insects. Darwin disagreed: 'There is no end to the adaptations . . . I fully believe that the structure of all irregular flowers is governed in relation to insects. Insects are the Lords of the floral world.'

In 1860 Darwin devised a technique that involved him poking a pencil point into the flowers of the early purple orchid and the pyramidal orchid to see what happened when an insect proboscis entered. The labellum, or the spring-loaded petal at the front of the flower, served as a landing platform for the insects. A slight touch to the fixed sexual organs freed the embedded pollen mass so that it became attached to the proboscis or the pencil by its sticky gland. Darwin then went on to work out the comparative anatomy of the flowers and how they became fertilized by different insect species. It was a wide-ranging example of adaptation by natural selection from a large range of complex flower species, the different parts of each flower fusing together or even reducing to almost nothing, the labellum and other petals varying in size and shape and altering the symmetry of the flower.

In June that year Darwin wrote to Hooker:

> You speak of adaptation being rarely visible, though present in plants. I have just recently been looking at the common *Orchis*, and I declare I think its adaptations in every part of the flower quite as beautiful and plain, or even more beautiful than in the Woodpecker.

He had been looking at the mechanism in the flowers which keeps the sticky glands fresh and sticky, and told Hooker that it 'beats almost everything in nature'. In the pyramidal orchid these sticky glands were

> united into a saddle-shaped organ [that] seizes hold of a bristle or proboscis in an admirable manner, and then another movement takes place in the pollen masses, by which they are beautifully adapted to leave pollen on the two lateral stigmatic surfaces. I never saw anything so beautiful.

Three years after the publication of the *Origin,* Darwin's explanation of these fast-evolving flowers was timely new evidence for his theories and the continuing interest showed how well *The Origin* had been received. The title of his 1862 book wouldn't sell so well today: *On the Various Contrivances by which British and Foreign Orchids are Fertilised by Insects and on the Good Effects of Intercrossing*. It set out the complete pollination process for several orchid species, some extending to spiders that were ambush predators of the insects that visited the flowers hoping for nectar.

Darwin showed that the orchids were much more devious than anyone had expected. A third of the species mimicked other plants that provided nectar and added to the deception with powerful odours that attracted particular species of wasps and other insects. The book did not however attract universal praise: 'if the Orchid-book had appeared before the *Origin* the author would have been canonized rather than anathematized by the natural theologians.' The review in the *Literary Churchman* found just one fault, that 'Mr Darwin's expression of admiration of the contrivances in orchids is too indirect a way of saying, "Oh Lord, how manifold are Thy works!".'

The discovery in the 1990s that some orchids select pollinators by smell would have amused the old friends at Down House, pleased Hooker and provided Darwin with yet another example of adaptation. His son Francis used to say that one of the things that made his father want to live a thousand years was to see the extinction of the bee orchid – inevitable, Darwin argued, because it had a self-fertilizing habit. In 1865, having 'almost given up' the search for insects cross-fertilizing the plants in certain seasons, he was still uncertain that insects pollinated all orchid species. He wrote that summer to a gardener called Traherne Moggridge:

> It is just possible that the same plant would throw up, at different seasons different flower-scapes, and the marked plants would serve as evidence. I conjecture that the Spider and Bee-orchids might be a crossing and self-fertile form of the same species.

Aware that the orchid Family has an unusually high number of species and knowing the ease with which they appeared able to evolve, he was preparing himself for what might become a sudden surprise. Had the Spider and Bee Orchids escaped from a larger population, like the hedge parsley escaping from the Sandwalk to the lane below the village? Two hundred years ago Goethe asked 'Who could blame us if we would refer to orchids as monstrous lilies?' Today, we would call them mutations. This concept of sudden change was not the anathema to Darwin that most historians have thought.

Breeders of orchids had always cultivated these easily changing flower structures, seeing their diversity as a symptom of vitality. The many colours and shapes are attractive features and their unpredictable expectations give gardeners a thrill, a sense of the unknown. Today, this is the basis of a thriving orchid industry with websites and society meetings to show off the latest variations. Yet as Darwin was searching his collection he was in pursuit not of beauty but monsters.

Instant monsters

Darwin had a knack of choosing good experimental animals and plants. Not only were orchids abundant at Cudham but they had more intricate structural features than most plants, more species and very elaborate mechanisms to stop inbreeding. Mendel's pea seeds were another good subject for experiments because it is easy to recognize either a smooth or a wrinkled surface and as mutants they were structurally explicable. Some of the variations in evening primrose, another subject of his experiments, were harder to explain, affecting many different features and even whole organs. In a lecture De Vries gave to the Royal Horticultural Society in 1895, he called these unexpected results 'hybridizing monstrosities' but no one could explain how they formed or might be inherited or where they fitted on Galton's polyhedron.

Although De Vries began his work with evening primroses and had no trouble in accepting that new varieties and species can only originate gradually, he was later forced to change this view. The switch of allegiance caused a lot of anger to many English biologists, especially A. C. Seward who in 1909 was organizing the centenary celebrations of Darwin's birth. Seward was a palaeontologist, professor of botany and vice-chancellor at the University of Cambridge. He was so faithful to the claims that evolutionary changes were gradual that he hardly accepted any changes at all. So he was embarrassed that his star guest, Hugo De Vries, had turned up with a new and opposing view, against the status quo and without much scientific evidence. Seward's ceremonial conference didn't encourage much new support for natural selection, even though genes were by then universally acknowledged to exist.

De Vries was much less isolated than Seward and was more aware of the social and scientific attitudes of European society before the First World War. This exposure to the shifting tectonic plates of European politics would affect his work. He argued that 'natural selection is a sieve, creating nothing as is so often assumed; it only sifts', and became a strong supporter of the sudden origin of species through mutation, declaring, 'I shall try to prove that sudden mutation is the normal way in which nature produces new species.' It was a change of mind partly influenced by the importance of quantitative science in the early twentieth century, influenced by Lord Kelvin's calculations for the age of the earth which offered insufficient time for Darwin's gradual processes to have happened. There must surely be another explanation for evolutionary change.

A few years earlier, Bateson had published a catalogue of structural abnormalities in which one body part was replaced suddenly by another. One was a mutant fly with legs on its head instead of antennae, and there were frogs and humans with extra vertebrae. Mutants like these were soon to become well known, but at the time there was no clear explanation of how they came about. They challenged the popular assumption that evolution happened gradually, but with little support they were not taken seriously by the majority.

In the 1930s another geneticist, Richard Goldschmidt, came up with an explanation for the origin of mutants, how evolutionary change happened and how structure was linked to ecology. He proposed that changes were determined by signals from a particular site on a particular chromosome. Goldschmidt was director of genetics at the Kaiser Wilhelm Institute for Biology in Berlin, where the students called him 'the Pope'. He preached sudden transformation, so that with a puff of smoke and a clap of thunder some new locus on the chromosome directly formed a new species. Goldschmidt's Jewish ancestry forced him to New York in the 1930s and in his lecture to the 1933 American Association for the Advancement of Science he christened Bateson's unfortunate flies 'hopeful monsters'. The mutations came from a sudden and quick change, controlled within the chromosomes and giving a structure that bred stable through further generations.

Goldschmidt's 1940 book *The Material Basis of Evolution* proposed a new system of life: 'evolution in single large steps on the basis of shifts in embryonic processes produced by one mutation'. In one example he showed how a breed of bow-legged dogs, the dachshunds, initially dismissed as mere monsters, were later cultivated to extract badgers from dens. Goldschmidt's ideas were furnished with other examples: gypsy moths with limbs on the wrong body segments; flatfish with both eyes routinely occurring on one side of the head; the joining of a bird's tail vertebrae to form a fan-like arrangement of feathers.

Initially no one took Goldschmidt seriously in America. The war in Europe paled his work into insignificance. Only now has he become an important scientist for the value he attached to the role of mutation in evolution, and he is fast becoming known as the father of a new multidisciplinary group interested in evolutionary development. In his theory, some mutations occur within a single generation, and the altered creatures that survive follow the pattern of Mendel's predicted ratios of inheritance. Modifications that happen in this way can become permanent features of the new species and manifest themselves on large and small scales. Galton's polyhedron did not rely on drastic events, therefore further evolution was erratic and unexpected.

Monsters at Kew

At Down House the orchids and primroses bloomed each spring and the garden gate turned rustier each winter. Not until the 1990s were the clues noticed, appropriately by the Keeper of Botany at London's Natural History Museum, Richard Bateman, who spotted mutant orchids with variations to the labellum, other petals and sepals. The changes are explained by different patterns of growth in the bud of the flowers during development. Inside the plant a peculiar gene controls the bud's development and its activity responds to the surrounding ecosystem and pollinating agent.

At about the same time as he found the strange orchids, Bateman also serendipitously found a mutant ginkgo tree in a private home just outside Kew Gardens. Cultivated plants of the maidenhair tree are often for sale in garden centres and they are planted on footpaths in towns and cities because they grow well in bad soil and smoky air. The tree at Kew often bore seeds not on distinct, separate shoots, but along the margins of the web-like leaves. As with the new orchids, this was explained by the unusual activity of a small gene that switched development processes in the early seedlings on and off. From Darwin's studies it was already known that the orchid flower was highly amenable to sudden changes of this sort, given its complex structure and involvement with so many pollinators and other dependents. But little was known about why the *Ginkgo* had mutated so rapidly.

Bateman is part of the new breed of general scientists driven by a love for sorting out evolving patterns in biology. He is part of a group of naturalists who are looking again at familiar plant mutants. Many of these changes turn out to be caused by chance through interactions on the genes from outside the environment. What made Bateman and his colleagues so excited were the clues that indicated that the orchid and ginkgo mutants were the work of interesting little genes very much like the ones that it had recently been discovered made dog breeds change so quickly and transformed the beaks of the finches that Darwin had found in the Galapágos Islands.

When looking for explanations of the ginkgo mutant bearing seeds on its leaves, Bateman's colleague Mike Frohlich found unusual activity in the cells at the growing tip of the short shoots. It was clear from the evidence that these strange specimens were hopeful monsters. The monsters appeared to have the same small gene found in several other plant species, ones from groups that have been around for many millions of years as well as very recently diversified orchid species. The team were then able to look at this small gene's DNA sequence. What they discovered was surprising: the gene seemed to have the function of driving other genes that control the growth of the flowers. Was this the gene that caused mutation or did it have a deeper purpose?

Evolution and development

In the 1980s, one of the first researchers to look for similar kinds of mutants was Ed Lewis at the California Institute of Technology. His area of study was the same kind of fruit fly mutants that had interested Bateson at Cambridge in the 1890s and Goldschmidt in the 1930s. Lewis was a student of one of T. H. Morgan's own students, making him a scientific grandson of that great fruit fly genetics pioneer. At Cal Tech, Lewis turned from embryology to wonder how some fruit fly mutants get an extra set of wings at particular body segments. He put it down to a small set of designer genes, ones specializing in wing development, acting on more than just their own body segment. They are analogous to the small controlling gene that Frohlich would later find in ginkgo.

Lewis was, according to his obituary in *Nature* magazine in 2006, a 'sweet, courteous and humble man' dismissive of mainstream scientists. He usually circulated his ideas as unauthorized copies, and he upset the US Atomic Energy Authority by suggesting that the nuclear explosions at Hiroshima and Nagasaki emitted radiation that caused cancer. In 1978, however, Lewis was persuaded to submit a paper to *Nature* about these small designer genes. It was a difficult paper to understand but it won him the Nobel Prize for physiology in 1995.

Throughout the 1980s, Lewis demonstrated his breakthrough to the general public with a set of coloured metal models. He set them out diagrammatically on a table top and, being very short, got under the table with a powerful magnet and began to move the shapes as an animated display of the small genes at work. In the 1990s the same kind of gene that Lewis found in flies was found in worms, mice, elephants and also in humans. The plant version of the similar kind of process accounts for Bateman's mutant orchids and ginkgos. They are called hox genes and they switch other genes in the embryo on and off, rather like the thermostat in your home controls your radiators.

Today, the hox gene is changing the way we think about evolution. Hox genes are small and few in number. Each body compartment or segment has up to ten of them, and they have retained much the same DNA composition since the major groups of animals and plants were first established. Being so ancient, these genes are universally distributed in modern animal and plant groups and have retained the same biochemical mechanisms of expression. By switching other genes on or off they control the development of all the body segments. While the hox genes stay the same through hundreds of millions of years the structural genes they control have varied through evolution.

By the end of the twentieth century large groups of scientists had found a single pattern to the way hox genes work. The mother's body releases special proteins once the egg is fertilized, and these block and unblock the hox genes at crucial times throughout the development of the embryo. Hox genes produce hox proteins at particular places in the embryo or in a bud and these eventually become the familiar body segments. The work of hox genes, switching the development of body parts on and off, depends on the ecology outside the cell. It's looking as though hox genes can explain how organisms adapt to the environment.

Only a few biologists can outline the essence of this complicated process: the concentration of hox proteins, their rate of production, how they switch other genes on and off in a growing embryo. It partly depends on the temperature, acidity and energy supply in the nearby cells and the outside environment. Particular combinations of these factors switch the biochemistry on and off in different ways, and control the production of each body part in the embryo. Hox gene mechanisms allow embryos to adapt their growth and development to particular conditions at a particular time, and this means that adaptations are unpredictable and happen by chance.

There are clear connections between these recent studies of hox genes by international research groups of scientists from different specialisms and the one or two early geneticists like Bateson and Goldschmidt, as well as breeders like De Vries and Darwin. The small hox genes control development within the conditions that happen to exist at a particular time. Similar genes explain hopeful monsters and the rapid adaptations and diversification of fruit trees, evening primroses and orchids. They programme development according to the state of the environment at the time, not according to any pre-planned design. For those still arguing about the validity of natural selection and the meaning of life, the existence of the hox mechanism means that there is no need to have intelligent design. It was what Darwin had thought as he sat in the summerhouse at the end of the Sandwalk, looking out at the beautiful Downe landscape: nature does it all on its own.

For a time in the 1990s it looked as though these advances meant that biology was well on its way to explaining how life works. From these indoor laboratories with their desktop theories it looked as though most of the evolutionary mechanisms were inside the cell. The molecular biologists had good DNA sequences fitting the latest structural details from extinct and living species, and those looking at how hox genes link developing embryos and their world outside were optimistic about explaining how things began. Some scientists, looking for smaller and smaller clues, were confident that the performance of hox genes during development had an ultimate role.

But when you go out into Darwin's garden and beyond into nature, looking for applications of these discoveries, the answers are not so clear. When you try to link the cellular processes to the ecological ones, to work out how the bits fit into the whole complex system, then there are still more questions than answers. Because parts of the system are always changing it is hard to see far ahead. In Darwin's garden it was as important to have binoculars as it was to have a microscope, let alone a timepiece.

12

Modern Ideas about Vertebrate Evolution

The 'Man book'

Early one morning in 1860 Emma and Charles were woken by shrill squeals coming from the pigeon loft. Etty's cat had got into one of the cages and had killed several of the most important experimental pigeons. Angered at the death of animals that were important to him, Darwin reacted quickly and with unusual force: he ordered that the cat be destroyed. Since her elder sister Annie's death several years earlier, Etty was a vulnerable child and when her cat died matters were made worse by her becoming seriously ill.

She never forgave her parents and perhaps by way of retribution Etty turned her interest in pets away from cats to dogs and her fox terrier Polly became a central figure in the life of the Darwin family. When Etty got married in 1871 it was her father who inherited Polly for they had already become very faithful friends. In his autobiography he called this 'love of dogs being then, and for a long time afterwards, a passion. Dogs seem to know this, for I was an adept in robbing their love from their masters.' The event that he was referring to happened when 'as a very little boy . . . I acted cruelly, for I beat a puppy, I believe, simply from enjoying the sense of power.' He said nothing more of the likely dependence of dogs and humans on one another or their mutual benefits from these entangled relationships.

So it was no great surprise when Etty heard from her mother that 'father is putting Polly into his *Man* book but I doubt whether I shall let it stand'. Emma was aiding her husband with the correction of the proofs of *The Descent of Man* and was trying to be helpful. She was afraid that the book told too much of human sentiment toward their pets, let alone detailing Polly's reproductive cycles alongside that of the apes. John Murray, Darwin's publisher, insisted on dumbing down the science and didn't even bother to consult the author about changing the title from 'Selection According to Sex', which had been Darwin's preferred version.

The new book gave further evidence that vertebrates were a single and complete branch in the Tree of Life. Man was not only related to monkeys but also to cats, dogs and every animal that had a spine. The branch included the whole range of modern vertebrate diversity including marsupials, placentals, reptiles and fish, and it went back in geological time millions of years before the dinosaurs originated. But although the vertebrates had many little side-branches, Darwin was confident that the group was a distinct entity that evolved forward, originating 'from some single progenitor. What its features were was impossible to speculate: something combining reptiles with mammalian characters.'

Darwin shared his ideas with his friends; similar proposals had been included in Huxley's 1863 book *Evidence as to Man's Place in Nature*, Lyell's *Antiquity of Man* (1863) and in 1869 Galton's *Hereditary Genius*. It was as though the X Club had agreed on a propaganda campaign to put humans in their place. For a time it worked and was supported by the scientific societies, but it was soon to hit an unexpected snag. The common thread of all these works show in their titles: humans were just another species on the tree of common descent, just another part of nature and with no supreme control or place within it. Darwin had summed it up clearly in *Descent*: 'Man still bears in his bodily frame the indelible stamps of his lowly origin.' Promotion of man's descent through natural selection stopped with these four books and it took several decades for more evidence and support to become available.

As the first supporters of natural selection by adaptation got older, there was no succession as would normally be expected from a new and hopeful intellectual movement. Biology was turning out to be a hotch-potch of so many different things: it was not like physics or chemistry with mathematical rules, nor did it offer a career path with money-making applications or a value to the military. Perhaps with the *Origin* biology might have reached a dead end. What was needed was new evidence to refresh the hunt for the key of life. They were all waiting for a breakthrough and none knew that it had already occurred in Brno.

The '*Man* book' was about the importance of sexual selection, as well as other human features such as language and consciousness. It was published in 1871 as *The Descent of Man and Selection in Relation to Sex*. Darwin was pleased to include the word 'Descent' rather than 'Ascent' in the title but was generally fearful of going public with the idea of demeaning the human condition: 'I shall be well abused,' he feared. Nevertheless, the book did become influential and was soon quoted in a House of Commons debate to argue for the question 'whether in each household the parents are cousins' to be included on the 1871 census form. Over the next decade Darwin would face grim criticism. One of his opponents began as a committed scientist with a promising career in anatomy. The other was a leading member of Gladstone's government and strongly believed that God planned nature's beauty.

The first of these, born in 1827, was a zoologist called St George Mivart. His anatomical work on carnivorous mammals and monkeys led to his election to the Royal Society's Fellowship and his campaign for this was helped by his close friend Huxley, who had been impressed by Mivart's style and skill in debating. Huxley had by then risen to the pinnacle of British science and was looking for bright supporters of the evolutionary cause to take the arguments into the next generation. 'As to natural selection, I accept it completely' Mivart told Darwin, but evolution by natural selection had become as much a political issue as a scientific one, a social cause as much as a religious one.

Mivart had a pedigree that stood him in good stead for his eventual admission to the X Club. He had been a pupil at Harrow School and then took up a pupilage at Lincoln's Inn. That led to him practising as a barrister and his inheritance of Mivart's Hotel, now called Claridge's. He had converted to Roman Catholicism when he was at Harrow and later this conflicted with the way he went on to think about evolution. As a student of Huxley's lectures, Mivart began to sense 'doubts and difficulties' with the new ideas. At least he had the courage to tell Huxley to his face that he was going to publish his carefully thought out criticism of Darwin's ideas. He described his reaction in a letter to a friend: 'As soon as I had made my meaning clear, [Huxley's] countenance became transformed as I had never seen it before. Yet he looked more sad and surprised than anything else.'

Mivart's criticisms were well considered and have taken until very recently to be answered in full. At the time they were the best scientific reaction to the arguments of natural selection, and Darwin knew it. That knowledge made it even more painful for him to read Mivart's provocatively titled response *On the Genesis of Species*. Without being able to do anything else with his anger he withdrew in his usual way and became ill. His continuing failure to find evidence for gemmules or obtain further support from other scientists for his theory of pangenesis offered no counter-offensives. The impasse was also a bad reaction for Mivart, because the absence of any scientific evidence to answer his own questions didn't encourage any real scientific debate about his proposals.

Mivart's case against natural selection was based on what lay behind his two important arguments. First, Mivart expressed disbelief that the 'placental dog' was such a remote relative from a 'marsupial wolf', or the modern horse *Equus* from the 10-million-year-old horse-like *Hipparion* that had roamed parts of Eurasia. His other contention was the difficult question of how complex organs such as wings and eyes could have formed without some long-term plan. He also used his legal skills to ridicule the concepts of gemmules and pangenesis, putting them down to Darwin's far-sightedness with no acceptable evidence. They were such early days in this very long investigation about the mystery of life that it was no wonder that the jury couldn't reach a verdict.

There was nothing for Mivart's opponents to say in constructive response. To his first challenge there wasn't the historical evidence from a fossil record and modern vertebrates didn't show any definite signs of linkage. To the second, an animal either flies or it doesn't and there's no halfway. The fossil record was no help either though the newly discovered specimen of *Archaeopteryx* was a slight challenge. One difficulty with missing links was that for every one found there were gaps for two more.

It was to take more than a century for scientists to provide sufficient evidence for satisfactory answers about these matters. Meanwhile everyone involved was going to have some disappointing part to play. Poor Mivart lost what he most desired: he was ostracized by the X Club for his negativity, and he was excommunicated from the Roman Catholic Church for his ideas about evolution two weeks before he died. Frustrated by the poor response to Mivart's challenges, *The Times* turned on *The Descent of Man* calling Darwin 'an old Ape with a hairy face'.

The second opponent of Darwin's concept of natural selection was Douglas Campbell, Duke of Argyll and Lord Privy Seal in the governments of Lord Palmerston and William Gladstone. He led the Victorian view that since Britain controlled the world it followed that man was responsible for what went on, put there as the guardian by God Himself. For most Victorian Creationists, it followed that any explanation of life's origin and diversification must involve God, but Argyll went one stage further and proposed that man had degenerated from an earlier perfect state at the Creation. It was the opposite to the popular view that early man had lived in 'utter barbarism' and has evolved to the present state, that he has somehow risen from animals. According to Argyll we had gone the other way.

When they first heard this account, members of the X Club

ridiculed Argyll. Huxley called him the 'Dukelet' and Wallace distrusted him: 'The worst of it is that there are no opponents left who know anything about natural history.' But one popular newspaper called Argyll 'The Darwinian Duke' and Darwin himself was always respectful of such an important representative of the aristocracy. This turned out to be an appropriate response because Argyll was a good listener and came to understand most of Darwin's arguments sympathetically. He accepted that 'apes are good in the wild' and that 'the subject can never go back to where it was before he wrote'. This led to the final accolade from Argyll when he served as a bearer of Darwin's coffin at the funeral in Westminster Abbey.

Perhaps Argyll had been thinking that catastrophes are an essential part of many different kinds of process – political, environmental and evolutionary. The British Reform Bills were trying to avoid revolution by accepting the forces of change through democracy. What better display of human force than that? But Argyll's criticism raised an important issue concerning the questions of the descent of man. If man evolved from some common ancestor, how could we uncover that history of human life?

Modern theories of descent

Theories about the origin and evolution of our own genus *Homo* have been hotly debated since Darwin's day and at last some consensus appears to be nearer than ever before. From DNA sequences of unusually well-preserved skulls, from the fossils and the tools of *H. heidelbergensis* and *H. sapiens* that came out of Africa, it is now clear that both species migrated to Europe. The first was the ancestor of *H. neanderthalensis*, which became extinct around 30,000 years ago. The second is us. Study of the fossils has proved that humans migrated and evolved according to the same constraints and patterns as any other species. There is no biological reason to think we are special.

In Darwin's time, vertebrate zoologists such as Owen and Cuvier used to spend a lot of their time looking directly at the skeletons of these fossils. Now there is a new breed of vertebrate scientist who put the DNA into a machine called a PCR-analyser and get a selection of amino acid sequences. They know that during development DNA will sense any changes in the environment, and the embryo cells will react and cause creatures to adapt accordingly. Molecular biologists have isolated hox genes controlling a small group of cells forming bone and found that they go on to modify the associated muscles, nerves and the vascular system.

These molecular geneticists have detailed genome sequences from the DNA of most living vertebrate species and some are being worked out too for the fossils. It was a big surprise in the last decade that these sequences are very similar from one species to another. In placental mammals more than 95 per cent of the DNA is the same in each species. The main differences come from the same genes switching on and off at different stages in development for each group. The idea that a standard plan for whole branches such as the vertebrates, controlled by that 95 per cent of the genome, could account for every vertebrate body now has a valid scientific explanation.

Owen was on the right track to expect some kind of programme to be inside the organism for we know now that it is the hox genes that switch on the supply of proteins during development. In mammals the hox genes perform these sensitive adjustments when the embryo is inside the mother. The system ensures a connection between the climate outside, the environment of the whole population, and the growing structural adaptations inside the body. The variety of these different conditions, however, is mind-boggling.

Darwin's strength was to look at as much evidence as he could from as many sources as possible, to perform experiments to test each variable separately and then to assess any progress. Most modern scientists follow the same method and nowadays, because there's so much knowledge and expertise available, they work together in interdisciplinary groups, often from several countries. With an exponential rise in publications from this kind of work there is a lot of evidence taking the nineteenth century theories of vertebrate evolution many stages further, confirming that most of those earlier naturalists started in the right direction.

Typical is work published in the science magazine *Nature* in 2007 by a group that compared 400 structural features of fossil and living mammal species and then compared them to different DNA sequences from living relatives. The results favoured an explosion in mammal variation 63 million years ago, from an isolated rabbit-like creature roaming the hinterland of central Asia, and leading to most of the placental mammals alive today. The 'Asian rabbit' was part of an even older lineage of small placentals that have been dated back more than 100 million years. This accumulation of knowledge offers a many faceted history with no clear centre. But all the evidence is slowly coming together and pointing to a common source.

Marsupials and finches

A particularly important consequence of Darwin's concept of a diverging evolutionary tree is that the same genes that occur in closely related living species will also have occurred in their common ancestors. That means modern DNA can provide unexpected snapshots from the geological past because some of those ancient genes survive for much longer periods of time than species; they remain in the system as records of earlier living processes and when the same sequences appear in several linking species, it means they are the oldest. Most species are thought to have been part of the living fauna and flora for about 7 million years, while many of their genes and smaller bits of DNA last for hundreds of millions of years in new combinations.

This idea opens up one of the most promising fields of study in evolutionary biology today and it is being applied to the topics that are most associated with Darwin himself: the marsupials and finches from his expedition to the southern hemisphere, the variations in breeds of dogs and pigeons that he loved so much at home, and the challenge of how the vertebrate eye works.

When he was in Australia Darwin had hunted some local mammals with a farmer and he proudly crated the prize platypus specimen back to Cambridge for study. The marsupials originated in North America and migrated south across Antarctica and east across to Europe. They began their migration, more than 100 million years ago in the Cretaceous Period, when they remained small and rat-like. Nevertheless they became victims of the competition with larger placental mammals and eventually became extinct in Africa, Europe and the Americas, each time being cut off as the land bridges connecting these moving continents broke away. Migrating from the danger of the more versatile mammals, they crossed Antarctica just in time, before it became too cold. The kangaroos and koalas became the largest species when they reached Australia though they gave Darwin 'very bad sport' when a farmer took him hunting.

Because the platypuses lay eggs, incubate them in nests and waddle around where the water is shallow, these strange crocodile-shaped creatures, which still breed in large numbers, were thought by some to be reptiles. The males produce venom through nasty spurs that they use to attack fellow suitors during the mating season. They are so unusual that in some northern hemisphere museums there was once a rumour that they were a hoax devised by some mischievous taxidermist. The females produce milk with the same kind of hormone control as other mammals yet it is sucked through teatless areolae on the surface of the abdomen.

In 1998 a group of Chinese and Swedish geneticists analysed some of the DNA and found that it revealed more similarity with the marsupials than the placental mammals. It was an unexpected result, different from what came from some fossil teeth which suggested that it evolved from the placental mammals. To try and resolve the question, an international group led by an oncologist called Randy Jirtle from Duke University, North Carolina, proposed to work out the whole platypus genome and compare it to that of the marsupial and placental mammals. The results are due to mark Darwin's two hundredth birthday.

There was an even better-known vertebrate group associated with Darwin's name but which, in fact, he did not even consider seriously, either on his world expedition or in his garden back home at Down. Yet the Galapágos finches are the group most often used to illustrate natural selection, perhaps because their soft and colourful features fit with Darwin's own kind and gentle image. In his 1845 *Journal* Darwin did acknowledge that on

> seeing this gradation and diversity of structure in one small, intimately related group of birds, one might really fancy that from an original paucity of birds in this archipelago, one species had been taken and modified for different ends.

Despite the popular perception, Darwin himself didn't do much with these specimens, other than to pack them carefully and send them off for storage.

What we know now about them came from the work of David Lack published in 1947 and resumed during the 1990s by Peter and Rosemary Grant and their daughters. They ringed and weighed and sampled the crops and measured the beaks. They did this regularly on living specimens through the seasons and compared their data with the changes in other neighbouring species. In 2006, with Arhat Abzhanov and his Harvard colleagues, they compared beak development to the activity of a gene that controlled the use of calcium in that part of the chick embryo.

There were clear differences in the diets of the fourteen or so species, and their beaks were adapted to the appropriate size and shape, causing their heads to have very different appearances. Three of the tree finch species ate insects and the one with the beak like a parrot ate fruit. The cactus finch had the biggest probing beak while the other ground finches that crushed seeds were of different sizes. It was discovered, therefore, that these species occupied ecological niches that were quite distinct because the precise size and shape of their beaks were crucial to their lifestyle and survival. The variation was due to the different expression of the calcium controlling gene. Changes showed up in the depth and width of the beaks during their development but the process took long enough for plenty of unknown factors to be involved as well.

Dogs and dinosaurs

A similar thing seems to be going on in the same body segment of dogs, from what may turn out to be the same set of hox genes that control the growth of the facial skeleton. It took several years work for two Texas medical biochemists, John Fondon and Harold Garner, to find the location of chains of duplicated hox genes in different breeds of dogs, and it appeared that these very short sequences of the DNA controlled the shape and size of bones that grew in the embryo. Comparisons with results from other laboratories showed that different breeds of dog had slight differences in their development due to these genes. Two of them, called Alx-4 and Runx-2, stopped growth before the wild type of adult wolf is formed. They were of crucial importance to understand how there were so many varieties of dogs, and how their species evolved. Breeders such as Fondon and Garner found that Alx-4 programmed the growth of the hind legs, and that Runx-2 controlled it in the highest vertebrae and the skull of each breed. But still these scientists didn't know the length and number of stages in the process, or why cats were so different.

Before these genetical links between development and the environment were known biologists couldn't explain how similar features showed up in distant branches of the vertebrate evolutionary tree. There was so much in common between a dog's head and a kangaroo's that even now it's hard to accept they have remote ancestry even though fossils and DNA comparisons agree that their lineages were separated more than 150 million years ago. We know that the Runx-2 gene is, and always has been, present in all these mammals, programming skull development in tune with their ecological requirements.

These hox genes are beginning to answer other difficult questions about natural selection which were asked throughout the last century and before by Mivart and others. An evolutionary tree of the visual features of vertebrates has been constructed based on the small variations of rhodopsin genes that control visual features in hundreds of vertebrate species. It is very similar to a branch of Darwin's original Tree of Life and scientists are now climbing back up to explore the ancestor's DNA. One laboratory that leads the study of these old genes is run by Belinda Chang at Toronto, where scientists have been trying to reconstruct the same rhodopsin enzyme that dinosaurs used to see in the dark 250 million years ago.

Chang's group began by spending a lot of time working out the base sequences of the DNA of that rhodopsin gene from hundreds of living species of land vertebrates. They checked and selected those parts which were most common in all the species, that is, the ones that have been around the longest. From these data they chose a final sequence and used this database to make their own artificial protein in the laboratory. Then they tested it in an actual chemical reaction by mixing it with cultured cells from a real vertebrate eye. The experiment worked: it is the same enzyme the dinosaurs used to see in the dark all those millions of years ago.

This kind of experiment takes modern DNA sequences back in geological time so that in the future even bigger databases from living vertebrates may lead to the oldest genes and these may help model ancient molecules. Another goal is to find out the form of the earliest vertebrates, a role most zoology textbooks still give to *Amphioxus*, the animal that looks like a tadpole and whose DNA increasingly appears like that of the starfishes. There is, however, so much else involved in the ancient processes that all this contemporary work does little more than scratch at the surface of one very large history. With some other groups the situation is even less clear.

There were more than enough of these clues in Darwin's garden to draw out features of the history of many groups, that is, the species comprising the branches of animals and plants. One came unexpectedly in the 1870s from Darwin's neighbour Sir John Lubbock, who had an interest in new optical devices that detected UV wavelengths of light. Using the equipment outside one evening, they found that ants move their pupae away from that danger into the dark. Only now is the full reason clear: that insects are particularly sensitive to UV light and use it for many different purposes. So do birds, but the whole branch of placental mammals do not. Accidentally, the neighbours had unearthed a useful clue about one evolutionary pattern. There were plenty of other pieces of evidence lurking in that garden, and we still don't know how many are still hidden there.

13

'A Most Perplexing Phenomenon'

Between two trees

Big, fat Betsy had been the parlour maid at Down House for many years and just before Darwin died in 1882 Emma pensioned her off with a small cottage and an income of ten shillings (50 pence) a week. It was a good deal for those times and the cottage had a pleasant view down the lane. But soon Betsy's health failed and she spent most of her time in bed downstairs, blocking off the draughty front door in that small room so that the only entrance to the cottage was through the back beside the pigsties and the stables of Down House itself.

Betsy's cottage was screened from Down House by a group of evergreen conifers and their dark green foliage made another sinister presence there, a barrier that was said to protect the house and lawns from the southwesterly gales that blew across from the Weald. More symbolically, it was the division between the poor on ten shillings a week and the rich of the big country house. The barrier became a meeting place where Darwin's grandchildren began to play with friends from the village and became their busy playground. They watched birds take the red fleshy seeds from the yews' branches and listened to their grandfather's stories of strange animals and plants on islands in the southern seas. The children enjoyed the swing between the two old yew trees, which moved them outwards and upwards into the open light, suddenly and powerfully.

Conifers like these yews and pines in the children's playground had been around since the Jurassic period, 150 to 200 million years ago. Their extinct ancestors stretched back another hundred million years and grew in warmer and very humid forests, together with large ferns and simple trees. Plants flourished in these misty swamps and inevitably some mutations formed simple seeds instead of spores in the female cones. It was a common trend in many of the populations in those forests and one that continues to happen on and off through geological time.

As recently as in 1970s Britain, fossils of these early trees were commonly found on the slag heaps of coal mines, the good specimens discovered there surprising botanists: the vision of the leaves of a fern bearing seeds several centimetres long was almost too fantastical for some naturalists. These seed-ferns were at their peak 300 million years ago and are seen by some experts to be the ancestor of many branches of seed plants including ginkgos, cycads, conifers and flowering plants.

Throughout the Jurassic, five major groups of seed-plants succeeded in becoming well established while many others introduced minor groups which only survived for a few million years and then became extinct. The early history of the flowering plants, however, was more difficult to discern. Slowly the old Orders of the two main kinds of plant life, the ferns, germinating from spores, and conifers, with seeds in cones, gave way to the more colourful diversity of flowering plants. But it would remain a mystery how one led to the other, which group of extinct seed-plants was the source of the flowers that cover most land on the planet today. What was it that distinguished the first flowering plants? Where did they grow and when? Why, once they had originated, did they suddenly adopt so many forms? The uncertainties continue and have even encouraged some to consider whether the flowering plants originated in some other way altogether, from mosses or large seaweeds at the same time that vertebrates began.

This abominable mystery

Darwin didn't take on these challenges until late in life, mainly because they were largely theoretical and he liked to experiment. As he got older this practical work in his garden diminished and he concentrated more on writing and reading inside the house. He still managed to take one good walk each day down the Sandwalk, resting in the summerhouse or in the potting shed behind the greenhouse where a fire was specially lit in the winter. His letters from those years show that he devoted a lot of his scientific thoughts to flowering plants and their physiology. He shared information about their migration and likely origin with some of the contemporary giants of biology, Louis Agassiz and Asa Gray in North America, respectively, professors of zoology and botany at Harvard University, Oswald Heer and the Marquis de Saporta in Europe.

From 1855 to 1882 Oswald Heer was professor of botany at the University of Zurich. He found many fossil leaves from the Alps and the Arctic in rocks that he thought were about 20 million years old. Now we know they are up to 55 million years old and show ancestry to the Order of the oaks and beeches, the Fagales. Heer suspected they were signs that ice had not always covered the Arctic or the Alps and that both regions had forests very similar to one another, and to the 55-million-year-old flora from the volcanic land bridge that linked what is now the Isle of Mull, the Faroes, Spitsbergen and Greenland.

Heer had described the species found in these fossil forests in 1875 and had surprised geologists such as Lyell with his suggestion that climate had changed from the times when the familiar white cliff deposits of the 'Upper Chalk' was formed. Now we know this happened from about 85 to 90 million years ago. Darwin had been impressed and had written to Heer saying so:

> The sudden appearance of so many Dicotyledons in the Upper Chalk appears to me a most perplexing phenomenon to all who believe in any form of evolution, especially to those who believe in extremely gradual evolution, to which view I know

> that you are strongly opposed. The presence of even one true Angiosperm in the Lower Chalk [deposited about 100 million years ago] makes me inclined to conjecture that plants of this great division must have been largely developed in some isolated area, whence owing to geographical changes, they at last succeeded in escaping, and spread quickly over the world . . . Many as have been the wonderful discoveries in Geology during the last half-century, I think none have exceeded in interest your results with respect to the plants which formerly existed in the Arctic regions. How I wish that similar collections could be made in the Southern hemisphere.

As usual, Darwin had put his finger on a part of the story that is still valid – the major rise in flower diversity between 100 and 85 million years ago and the isolated location of the earliest records of the earlier newly evolved flowering plants. After Hooker's expedition to Australasia and India in 1870 Darwin had begun looking for these earliest ancestors, expecting them to be confined to small islands in the southern hemisphere. The letter to Heer went on to share memories of their mutual friend Charles Lyell, who had just died, and ended with a curious reminder of how new technology has always fascinated the scientific mind:

> The death of Sir C. Lyell is a great loss to science, but I do not think to himself, for it was scarcely possible that he could have retained his mental powers, and he would have suffered dreadfully from their loss. The last time I saw him he was speaking with the most lively interest about his last visit to you, and I was grieved to hear from him a very poor account of your health. I have been working for some time on a special subject, namely insectivorous plants. I am very much obliged for your photograph, and enclose one of myself.

Although it was now over fifteen years since publication of the *Origin,* Heer was typical of many in central Europe who still had their own peculiar way of reconciling evolution with Creationism. In his reply to Darwin, Heer wrote:

> Although a species may deviate into various forms, it nevertheless moves within a definitely appointed circle, and preserves its character with wonderful tenacity during thousands of years and innumerable generations, and under the most varied external conditions . . . Even if the first species were extremely simple, for them an act of creation must be admitted . . . Great creative renewals are indicated within the limits of the principal geological periods; and during those periods important transformations also took place.

But in Aix en Provence, a palaeontologist called the Marquis de Saporta had a more progressive explanation and in 1877 Darwin wrote to him with some excitement:

> Your idea that [flowering] plants were not developed in force until sucking insects had been evolved seems to me a splendid one. I am surprised that the idea never occurred to me, but this is always the case when one first hears a new and simple explanation of some mysterious phenomenon.

The point was well made and the sudden expansion of diversity in insect species at the same time as flowering plants clearly showed the dependence of these two very large groups on one another. It was a dependence that reached the highest effectiveness with the orchids and insects making use of one another in thousands of different ways. 'The beauty of flowers, their sweet odour and copious nectar, may be attributed to the existence of flower-haunting insects, but your idea goes much further and is much more important,' Darwin wrote to de Saporta. 'Such animals as deer, cows, horses, &c. could not flourish if fed exclusively on the graminae and other anemophilous monocotyledons.' The observations confirmed Darwin's beliefs that the 'suggestion of studying the manner of fertilization of the surviving members of the most ancient forms of the dicotyledons is a very good one'.

The next year Darwin read more of the topical work on flowering plants and was particularly intrigued with an essay by the geographer John Ball on the origin of the flora of the Alps. In it, Ball offered an ill-conceived theory that the altitude of flowering plants was dependent on carbon dioxide concentration in the atmosphere. Darwin wrote about it to Hooker in July 1879:

> It is pretty bold. The rapid development as far as we can judge of all the higher plants within recent geological times is an abominable mystery. Certainly it would be a great step if we could believe that the higher plants at first could live only at a high level; but until it is experimentally [proved] that Cycadeae, ferns, etc., can withstand much more carbonic acid than the higher plants, the hypothesis seems to me far too rash.

Then he moved to more agreeable theories:

> Saporta believes that there was an astonishingly rapid development of the high plants, as soon [as] flower-frequenting insects were developed and favoured intercrossing. I should like to see this whole problem solved. I have fancied that perhaps there was during long ages a small isolated continent in the S. Hemisphere which served as the birthplace of the higher plants – but this is a wretchedly poor conjecture. It is odd that Ball does not allude to the obvious fact that there must have been alpine plants before the Glacial period, many of which would have returned to the mountains after the Glacial period, when the climate again became warm. Ball ought also to have considered the alpine insects common to the Arctic regions. I do not know how it may be with you, but my faith in the glacial migration is not at all shaken.

It is still a big question why the flowering plants originated so much later than all other major groups of animals and plants. All animal Phyla were established more than 500 million years ago, the ferns and mosses 300 million years ago, yet the oldest fossil flowering plant is less than half that age. Another big question not yet answered is which group or groups they came from.

Since Darwin first raised these issues with Heer, de Saporta and Hooker, different generations of palaeontologists have tried to provide answers.

Not long into the twentieth century more details of flowering plant behaviour were discovered. The most important feature was that two sperms are needed for fertilization. Usually insects or wind disperse the male pollen and it sticks to the female carpel. The sensation of these organs touching one another sends out a pollen tube to the egg and two sperms swim down into it. One penetrates the egg, causing the cells to divide into an embryo, the other fuses with nearby cells to form nutrients in the new seed's fruit, a kind of 'placenta'. The need for two sperms is a character that the flowering plants shared with only a few other groups such as ginkgo. The conifers need only one sperm and share that character with just a few of the first angiosperms, their living relatives such as water lilies.

In 1990 there was a surprise discovery that two sperms were needed for fertilization in some other living species of conifer-like shrubs that have always been awkward to classify and place in an evolutionary sequence. They were called *Ephedra* and *Gnetum* and have been seen as at the boundary between flowering and non-flowering seed-plants since Darwin's day. The revelations showed the feature of fertilization by two sperms to be present in two species of real conifers as well. These unusual examples match another group of unusual non-flowering seed-plants and another strange method of fertilization. Ginkgo, cycads and some extinct conifer-like plants have sperms that swim free during fertilization, when they leave the pollen tube that falls short of the egg.

Evidence of early angiosperms

Goethe was the first to devise a theory of origin based on structural features that are exclusive to flowering plants, rather than as part of any evolutionary scheme involving other species. It was a view being reconsidered in places like Cambridge at the start of the twentieth century as a reaction to Darwin's very holistic natural selection. It was there that Albert Seward was trying to regain some stability within evolutionary thought while William Bateson was causing difficulty for Darwin's legacy. There was also the husband-and-wife botanists, the Arbers, who suggested that flowers were little more than specially adapted leaves. The ardent fossil hunter Newell Arber found several specimens with sterile leaf-like organs close to the seeds of an extinct group of Jurassic fossils. He thought that they were getting close to being real flowers, with at least the remains of stamens in an adapted fertile male leaf.

When her young husband died in 1918, Agnes Arber stayed on in their little house on the Huntingdon Road and converted one of the bedrooms into a laboratory. Women were not allowed in many of the university buildings in those days and she flourished as an independent spirit, eccentric and kind. Her tutorials were very popular and she became a respected tutor at Newnham College. Gwen Raverat, Darwin's granddaughter, and Virginia Woolf were among the many students she influenced with her hopes to link science and art, and Woolf's 1929 essay about 'a room of one's own' was very likely inspired by Arber's little laboratory.

With her geologist daughter Muriel, Agnes dedicated her life to teaching the importance of looking at the whole organism and its environment for understanding evolution, not just the analytical mechanisms that were beginning to take over in the male territories such as laboratories. She upset her increasingly objective colleagues by using words like 'urge', 'endeavour' and 'perseverance' to describe a plant's actions and said they were 'unavoidable because we have no other set of terms in which to express that compulsiveness by which the plant works'. The full expression of her conclusions, botanical and philosophical, was finally made in 1950, in a book with a title that challenged humanity – *The Natural Philosophy of Plant Form*: 'The leaf is a partial shoot, arising laterally from a parent whole shoot, which has an inherent urge towards the development of whole shoot characters.' Her concept of how leaf adaptation might give rise to flower parts was explained with feeling: 'In the vine a tendril sometimes occurs in place of a bunch of grapes, and from this the author concludes that the tendril is to be interpreted as a bunch of grapes incompletely developed.'

Yet the tide of biology was against Arber. Through the twentieth century scientific appraisal considered smaller details, led by technology splitting organisms up into finer parts and eventually putting them back together again. It had become a universal truth that the holy grail of evolution would be found by investigating further and further into the body of the organism in order to find the secrets of natural selection. Inevitably these very large databases got sorted statistically and the specialists from different disciplines argued about which evolutionary tree that emerged made best sense. The physical and biological scientists who tackled these problems usually set one view against another. The investigations revealed many details but no single answer.

One of the difficulties that many encountered was the embarrassing truth that nobody knew for sure what a flowering plant actually was. With the regular improvement of new technologies early hopes of answering that question were dashed each time new techniques gave new information that invariably led to new definitions. More powerful microscopes, new dating methods for the fossils, statistical analysis of databases and most recently automatic recognition of DNA base sequences have all offered much but delivered no answers.

In addition to this new technology, new discoveries of fossil and living specimens forced scientists to return to the question. Early flowers and fossil pollen grains with their distinctive ornamentation have been found in sediments from many places older than the Lower Chalk of the North Downs. In the 1960s the confirmation of continental drift as a viable drive for animal and plant migration and separation was a great boon to Darwin's ideas on divergence and this set palaeontologists looking for the geographical centres of origin as well as structural features.

In the 1970s the scanning electron microscope helped studies of pollen and, from the 1980s onwards, of tiny fossil flowers more than 100 million years old, mostly from southern Portugal by the Scandinavian botanist Else-Marie Friis. In the 1990s further fossil flowers were found in older rocks in China and more recently from Liaoning Province a new fossil called *Archaefructus* which may be around 124 million years old. It has no petals and sepals, pairs of stamens just below the carpels, and thin stems and dissected leaves suggest the plants were aquatic.

In 2006 the reappraisal of a shrub living on the isolated South Pacific island of New Caledonia made a lot of botanists very excited. That was because they thought it was a last relic of one of the earliest flowering plant families. It was named *Amborella* and has tiny greenish-yellow flowers and red fruit, and its egg cell is fertilized by just one sperm, not by the usual double fertilization process of all other flowering plants. There are other woody species such as the magnolias, tulip trees and *Ascarina* which share many of the features of these first flowers, and many of them are still growing in the southern hemisphere. Other botanists have been influenced by clues from living freshwater plants such as *Chloranthus* and *Hydatella*, whose many simple features, wet growing conditions and southern locations made them likely candidates as ancestors of the first flowering plants.

The whole state of the order, names and shape of the Tree of Life was further scattered into disarray by work going on in the laboratories that were analysing DNA in studies of evolution and taxonomy. Mark Chase is the leader of a team at Kew Gardens and has spent several years checking and double-checking the sequences of nucleotide bases. It is a very small proportion of all the DNA inside one plant cell, but it is stable and easy to extract and purify. One big risk that Chase and his colleagues worldwide are taking is to give great attention to just three small pieces of DNA, although they are from very different parts of the cell.

In conjunction with other research teams, Chase and others formed an interdisciplinary data-sharing forum which had different ways of storing and analysing the molecular and morphological data. They first put online a classification for the Orders and Families of flowering plants in 2003. It was a new and growing version of the Tree of Life comprising what some call the gene tree for plants. This was based on consensus between results of statistical analysis of all sorts of data: visual, microscopic and chemical.

A common view was that their tree showed both useful classification, which would have pleased Hooker, and a good evolutionary path, which would have pleased Darwin. Even so, many botanists were surprised to read on the website that several Families of flowering plants had been given a different ancestry. The molecular evidence suggested a very early divergence of a single evolutionary branch for all the flowering plants, in comparison to the structural evidence which suggested a number of different extinct conifer-like seed plants as favoured ancestors. The evidence leaves profound uncertainty about which flowering plants, extinct or extant, were more closely related to the earlier seed-bearing ancestors.

A need for some lateral thinking

It is still universally assumed that flowering plants had one common origin. It follows that it must have been in one place at one time so it is a heresy to suggest that there was more than one place or time of origin. Equally unacceptable must be any seed plant species that have arisen in different places from different sources. Such speculation is in conflict with some of Darwin's fundamental thinking. Yet these traditions should be challenged from time to time especially when progress gets stuck.

So many bits of evidence have accumulated from the new molecular evidence and from the newly discovered genera of fossils and living flowering plants that it is hard to put them all together, especially since most of them are separated by geographical space and geological time. There were enough facts to suggest that the flowering plants may have come from many extinct groups including conifer-like plants, ginkgo-like plants, Arber's extinct Jurassic seed plants, old cycads, seed ferns and a few others. The new DNA sequences, however, don't support any particular one of these possibilities. Nonetheless, molecular biologists have offered evidence that hox genes can drive the sensitive processes inside each embryo cell during early development, each body segment being programmed separately and each being vulnerable to sudden changes. The resulting mutations gave hopeful monsters that usually died but sometimes led to more efficient individuals. The usual flower structure is determined by hox genes and has a particular number of sepals, petals, stamens and carpels for each species. The few extremes, such as *Archaefructus*, which has no sepals or petals, are from the mysterious no-man's-land in between the established groups.

Another discovery is that at certain moments the chromosomes in several flowering plants doubled and sometimes tripled. Such increases in the number of genes, spurred by actual events which may have coincided with times of major environmental change, may be accounted for by sudden switches of their hox-like genes. Certainly, there was one such gene in early seed plants, called *needly*, which specified the female spores. The ABC theory of development is the notion that there were three comparable genes that controlled the way a flower developed: gene A formed the sepals, genes B + A the petals, genes B + C the stamens and C the carpels.

Needly is still present in modern conifers, performing the same function. It can, however, go wrong, and confer the female spores on a male shoot. One scenario is that this was surrounded by a layer of protecting tissue to form a real carpel and give the conifer-like plant a real flower. The idea can be tested by searching for the gene in the DNA sequences of early flowering plants and their early conifer-like ancestors, and this is what Mike Frohlich of the Natural History Museum is investigating now.

Organisms are always looking for opportunities and ways to change, whether suddenly with chromosome duplication or mutation, or more slowly with only small biochemical differences. The hox processes in the embryos created the mutations that suddenly began the new seed groups from the conifers and possibly other early seed plants. These involved the establishment of processes like fertilization with two sperms and the adaptation of flower parts by special hox gene instructions to very particular buds of developing leaves.

Every time there is a discovery of a new technique, a really old fossil flower or a peculiar living plant, scientists rush to investigate. When these alerts are taken together a picture of the source of flowering plants begins to form, but what happened is still very unclear. Perhaps these biologists are looking for something that never happened. As Agnes Arber might have said: with perseverance, can some new endeavour explain an urge, within several groups, to adapt in different places at different times?

'I ask you to look back, not on a proud record of the success of famous men, but on an unbroken record of failure.' This was not a comment on the traditions of the X Club but on how scientists understood the early evolution of the flowering plants. They were the opening words of the palaeontologist Tom Harris in his address to the British Association for the Advancement of Science in 1960. But perhaps, instead, the history of our researches into the origin of flowering plants is a record of a devious process that cannot be measured by the objective methods of modern science. Perhaps we will find the answer in future, perhaps not. Perhaps we are looking in the wrong place asking the wrong questions.

14

Origins of Life

The first big catastrophe: bugs from a warm pond

The young son of the squire at Downe, also called John Lubbock, helped out on many of the Down House projects, and he liked especially to go pond-dipping in the pond on the village green. He took jam jars full of the green water back to examine under his new brass microscope. It magnified so that specimens of the minute transparent crustacean *Daphnia* showed the heart and guts as well as a segmented body of abdomen, thorax, mandibles and antennae.

In the middle of the nineteenth century a lot of new detail was being discovered about the anatomical and cellular structure of animals and plants. Some groups, like the corals and barnacles, were more difficult to discern than others. And the scientists among the naturalists were looking for evidence from the fossils as well as living species to test the new possibilities of evolutionary change. The difficult groups had features that changed slowly and evenly through time and showed nothing that gave any hint that sudden catastrophes influenced variation. Their quietness fitted the idyllic ambience of Lyell's uniform gradualism.

The French had tended to lay aside much of their interest in evolution once Darwin's work was published. He had, after all, disproved what most of their great predecessors – Lamarck, Cuvier and Geoffroy – had so famously elucidated. Instead, through the 1860s, the French intellectuals were occupied with the breakthroughs in biological chemistry led by Louis Pasteur. His debunking of spontaneous generation just after publication of the *Origin* took French attention even further away from the revelations at Down House. Instead, they saw Pasteur's demonstration that life doesn't arise spontaneously as a very serious challenge to Darwin. Implicit in Darwin's theory of natural selection was the notion that life, at some point, originated from no life.

But Darwin was interested in what was going on in Paris, later writing in 1871 from Down House to Joseph Hooker at Kew:

> If (and oh! what a big if!) we could conceive in some warm little pond, with all sorts of ammonia and phosphoric salts, light, heat, electricity etc. that a proteine [sic] compound was chemically formed ready to undergo still more complex changes; at the present day such matter would be instantly devoured or absorbed, which would not have been the case before living creatures were formed.

There were, however, too many uncertainties in his thinking that kept pulling Charles back into the safety and routine of family life at Down House and he had no actual evidence to support ideas of how organic life may have begun. All their speculation lacked a first cause. The pond was a wonderful metaphor, but it was fundamental evidence that was required in these early days of science and technology.

On the other hand, Darwin was confident that his scientific view of how the diversity of species had come into existence was on the right lines. Any dialogue about it was difficult with those he loved, like Emma, and he was troubled by the reaction from some academic partners. Lyell, for example, was not only God-fearing but he was also sceptical about natural selection as a process, especially if it had to work quickly. Darwin secretly dreaded his friends' increasingly frequent visits to Down House as the 1860s progressed.

One exception was Ernst Haeckel, the professor of biology at Jena, who had interests in the links between inorganic chemistry and organic biology. He was thinking seriously about catastrophe as a device in evolution and had observed sudden and unexpected changes during the development of several vertebrates. He set out his interpretations in his 1866 book *Generelle Morphologie der Organismen* and visited Down House just after its publication. 'Your boldness sometimes makes me tremble, but as Huxley remarked, someone must be bold enough to make a beginning in drawing up tables of descent,' his host at Down House warmly remarked in appreciation.

Later during his visit, Sir John Lubbock joined them at dinner. At the table discussion quickly turned to Haeckel's book, which suggested lots of parallels between embryology and evolution, and that many developing embryos reflect progressive stages in the evolutionary succession, especially of vertebrates. Haeckel claimed the stages that an individual takes during its development in the womb or egg follow the evolutionary lineage of that individual's species: so the embryo starts as a cell and then develops in clearly programmed stages. The dinner party went well, Emma noting 'I have seldom seen a more pleasant, cordial and frank man – a great good-natured boy.' The respect was mutual, for Haeckel was impressed by Darwin's 'broad shoulders of an Atlas that bore the world of thought: a jovial forehead as we see in Goethe'.

Haekel's ideas inspired others and in the 1870s another German biologist, Wilhelm Schimper, started work on a biological equivalent to Pasteur's experiments, exploring how cells are generated. His observations under the light microscope in 1885 showed green particles inside many plant cells that always divided independently. These were chlorophyll-containing chloropasts, the sites of photosynthesis. In 1905 Konstantine Merezhkovsky, professor of botany at Kazan, suggested that chloroplasts made their own protein and handed down their characters without the involvement of a nucleus. It was as though they were separate very small single-celled organisms living inside these much bigger plant cells. He claimed 'they are symbionts, not organs', meaning that the formerly separate organisms had become dependent on one another, each providing something the other needed. It was the first suggestion that chloroplasts originated as bacteria.

Merezhkovsky's discovery led on to a mysterious relationship between two scientists from very different backgrounds but with the same idea of how life began. Alex Oparin, a biochemist from near Kazan, in 1924 formulated the theory that life originated in a primeval soup. Also in 1924, J. B. S. Haldane published the same theory about life's origin, it is said, independently. Neither Oparin nor Haldane had any actual evidence from palaeontology or genetics, though Haldane lived to see experiments that synthesized organic molecules from simple inorganic ones as well as geological evidence of an early atmosphere without oxygen. Their philosophical arguments have since been overlain by other advances in geology and molecular biology. Today, the latest explanations of how early life became established reject a lot of the intervening assumptions.

> I've been put here to pick up where Darwin left off! All of his descendants dropped the ball and you had to turn to some people who knew about molecular biology in order to get the ball rolling again. But now it's happening, and I'm feeling more hopeful than ever.

So declared the colourful and eccentric microbiologist Carl Woese (pronounced 'woes') in 2002 about how life preceded the emergence of multicellular organisms. He picked up the ball with results he obtained by sequencing the single helix of RNA, half of a double helix of DNA. The data came from the kind of bugs that he thought had been around 3.5 billion years ago and had survived up to a well-defined event during the Cambrian a mere 0.5 billion years ago. He called them the Archaea, and argued that their 'microbial world accounts for 95% of the diversity of life on this planet'.

In a photograph for a student newspaper Woese, in his red-and-black checked shirt, sits back on his swivel chair, with his Nike-clad feet on the desk beside the computer screen. He was being interviewed by the student newspaper:

> Let's face it, I'm a pretty strange guy, and it's nice to loosen up. Beer helps me to unwind. I don't get drunk like you do at fraternity parties – I just sit there and get mellow and see the world in a few more dimensions!

This small-boned slender man sounds very unlike Charles Darwin in how he describes living but he is having a big impact on how we think about the origin of life.

Archaea were the first organisms, methane-dependent single cells living in hot, harsh environments 3 billion years ago. Now they're restricted to sulphur springs and seabed volcano vents. Most have a single circular molecule DNA and their RNA has more repeated sequences than in other organisms. Woese thinks that some of this comes from a process in which bits of RNA come together inside some simple cells. The thinking is a stage further on from the scenario envisaged by Oparin and Haldane that involved the alluring image of a primordial soup and lightning strikes.

From his molecular evidence, Woese believes there are just three life forms that have evolved separately even though they regularly exchange genes. As well as the Archaea there are bacteria and nucleate cells, leading to the familiar biology of fungi, animals and plants. Woese goes on to challenge many of the established biological conventions. 'The time has come for biology to go beyond the Doctrine of Common Descent', he said in the interview. Instead, he offers a communal environment with only loosely organized cells exchanging genes and cell contents according to the particular environmental influences. 'We cannot expect to explain cellular evolution if we stay locked in the classical Darwinian mode of thinking.' In 2006, Carl Woese was elected to the fellowship of the Royal Society, a sure sign that his revolutionary ideas had been accepted.

It was not such a far shift from the image of Galton and Darwin walking along the pebbled Sandwalk in the garden at Down House. Regularly, they had looked down at the complex system of the pebbles set in the sand, gathering data by counting the stones and looking for patterns in the plants. Most of the options in their theories came in ignorance of the detail that was involved but they all boiled down to a single idea, selection of one thing or another. While Darwin and Galton were pluralists and could see many of these options together, most of their modern successors are specialists who see only one thing at once, usually focusing on a much smaller scale.

Six hundred million years ago the diversity of species was much simpler for there were only unicellular organisms. Suddenly, from ten to fifty million years later, living systems experimented with groups of cells forming multicellular organisms. At the same time the amount of oxygen in the atmosphere rose dramatically. On the other side of the same equation, the level of carbon dioxide in the atmosphere fell through the same period of time.

Very few specialists dispute the thesis that photosynthesis oxygenated the atmosphere, first from bacteria and later from plants, and that this was the principal source of the organic carbon which got buried in the deep ocean and which we are now burning off to increase atmospheric carbon dioxide once again. With the increase in the number of cells, photosynthetic organisms diversified and atmospheric oxygen increased. The first organisms to become well established in the plentiful shallow water environments were the sponges and just a few million years later these led to the radially symmetrical polyps and anemones. It was how more complex life was born.

Since Woese's work became well known others have agreed that these changes in the Cambrian, more than 500 million years ago, involved the movement of stray bits of RNA into the cells of the early invertebrates. Similarly, the transfusion of bacteria into plants was thought to be limited to just a few species but more bacterial groups are being reported as engaged in inter-kingdom genetic exchange. You see some examples of the union in the tumour-like outgrowths on the old oak trees that Emma and Charles planted in the copse. They are growths caused by a pathogenic bacterium that changes hormone concentration in the host so it loses control of the rate of cell division. Woese suggested that this may be the kind of relationship that happened between bacteria and other groups back in the Cambrian.

New animals after an iceball earth

No doubt the geologists Paul Hoffman and Dan Schrag were far away from Woese's fine theories of early life when they were burning the midnight oil at Harvard in the cold winter of 1998. They were also unaware of the many details of zoology and botany that may have had some remote connection with the very old environments they were modelling. They were certainly unsure of the theoretical considerations of convergence and morphological design, the meteorological factors of atmospheric pressure, the special requirements of small molecules of RNA.

Instead they were experienced in taking data from the chemistry of rocks and atmospheres, from the distribution of continents and oceans, and writing computer programs to select the best fit to the theories they were testing. The one argument that interested them most concerned the changes in surface temperatures of the whole planet 550 million years ago. But they had to keep all these other things in their minds, even though they were not that kind of specialist.

One night, however, Hoffman and Schrag came up with a theory called snowball earth to explain the sudden increase in biodiversity. It required high concentrations of carbon dioxide, 300 times more than in today's atmosphere, that reduced the earth's surface temperatures to below freezing. The models varied slightly and in some the planet was a slushball rather than a snowball, and maybe water at the equator was not frozen at all. The different models explored different scenarios and there was not enough evidence to know which was right. But what did seem clear was that for the planet to lose its surface ice the model also required a warm period to end that ice age; perhaps even a very hot period by the standards of the earth's temperature since then. There was little evidence to support this suspicion but whatever happened it was on a big scale and it lasted for millions of years.

Could the sudden release from these extreme conditions have given the organisms that survived the great freeze an opportunity to radiate sufficiently to produce what some palaeontologists have called the Cambrian Explosion, just over 500 million years ago? For 10 million years after these extremes there was a long slow recovery in the number of living organisms in the warm water. We know this from the great range in the animal structures that evolved through this short interval. A major source of evidence was first discovered during the First World War by an American palaeontologist Charles Walcott who stumbled on the now famous spot in the Burgess Pass, going through the Rocky Mountains in British Columbia. The locality has a huge variety of well-preserved and strangely shaped fossil animals. Walcott collected thousands of specimens, now housed at the Smithsonian Institution in Washington, DC, and they have been poured over by generations of students ever since. Even the difficult Burgess Shale specimens are now finding a place in familiar Phyla, which only a few decades ago would have been unheard of.

The detective work involved in finding evidence of a recognizable group within Walcott's specimens demands the kind of painstaking reconstruction that you'd expect at the end of an Agatha Christie mystery. One recently discovered fossil, now housed in the Royal Ontario Museum, Toronto, is what turns out to be a mollusc. The fossil's mouth has teeth positioned in rows on a ribbon and when the first row becomes worn they drop out or are swallowed. Then the ribbon moves forward with new teeth. The same thing happens to the snails you see in the microbial film inside a glass fish tank. The teeth that had dropped out had been found separately and weren't associated with the other remains until recently.

Many of the structures seen in these different animal groups are variations of the same features, examples of what is known as convergent evolution. More than 150 years ago Richard Owen recognized the recurring features in vertebrates when he compared fish fins and tetrapod limbs. It was also what worried Mivart about natural selection in the 1870s, common themes in animal and plant form that Darwin never could explain. Now we know it is the kind of modularity that comes from the function of genes, which work to adapt these segments in modern animals as they would have done in the very early organisms of the Burgess Shale. Their role of changing the structure according to the outside conditions ensured that animals adapted an efficient bilateral or symmetrical structure and then went on to evolve segments from a head to an anus.

Because there is little detail about this early life there has to be a lot of speculation. Recent finds include a worm-like ancestor dated to 550 million years ago. All else we know for sure is from its prints left in the mud, just a few millimetres in diameter. This organism differentiated five to ten regions from its head to tail and a few more sideways, from its back to its belly, giving a kind of chessboard organization. This included a heart to serve as a pump, light-receptor cells and a central nervous system. The chessboard pattern allowed this single form to adapt each part for different environments, places with different light regimes, different sensory effects, different temperatures.

Some of this worm-like ancestor's descendants became bilateral animals such as slugs, worms, insects and fish-like early vertebrates. Others were radially symmetrical and became animals like sea urchins. In all, about thirty different body plans were established and they have all diversified into the Phyla that comprise the animal Kingdom. As far as we know, none has become extinct. All the main animal groups came from this worm-like ancestor with the all-too-familiar body plan.

Self-organized order

Francis Galton was convinced that a lot of these complex systems organized themselves, and to demonstrate how they worked he began measuring their rates of change. In his eccentricity, at a tea party in Down House, he was persuaded to investigate how tea was brewed. Another time he looked at changes in the weather. His equations included measurements of the size and temperature of the teapot and the weight of water used. His maps of weather systems showed the direction a cyclone rotated, where the air moved upwards, as well as temperature and pressure isobars. Each appeared to be a process drifting aimlessly in a different environment and getting on very well by chance.

But the worms and other unseen animals and plants that lived through the last 550 million years of geological time were far from drifters. There was method in their apparently random madness, and the patterns they followed are now known to be those first recognized in Darwin's garden by Galton, who had sensed their importance while reading the work of a French physiologist, Claude Bernard. During one of his visits to Darwin's garden, Galton joined in a family game they called 'Fools' Experiments'. The prize was for the first to see the earthworms Charles had set out in jam jars respond to different musical rhythms. It was a game Galton enjoyed because it attempted to measure the seemingly unmeasurable. Emma played loudly on the piano indoors, Francis played the bassoon and Darwin whistled. Emma smiled that they had 'taken to training earthworms' but didn't

Body plan of the hypothetical ancestor to bilateral animal phyla. Burrows in Precambrian sediment suggest they may have been 0.5cm in length. The same regulating hox genes are present in all subsequently derived phyla.

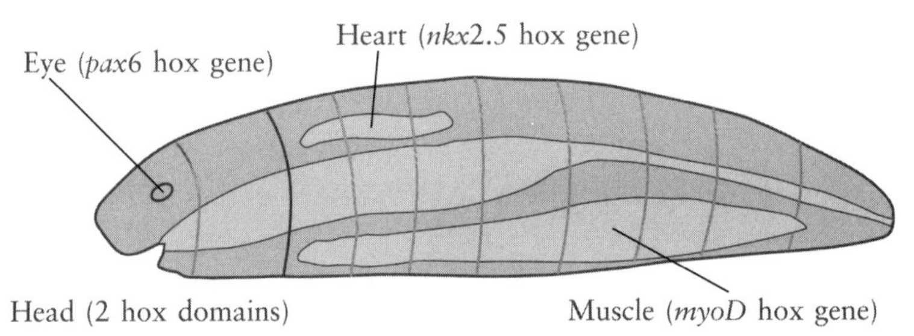

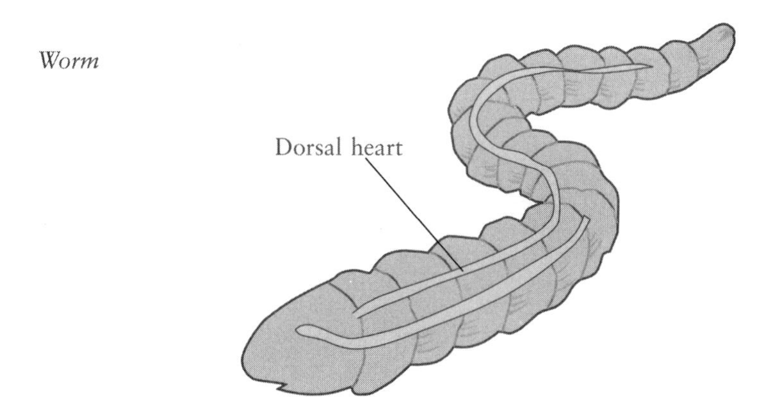

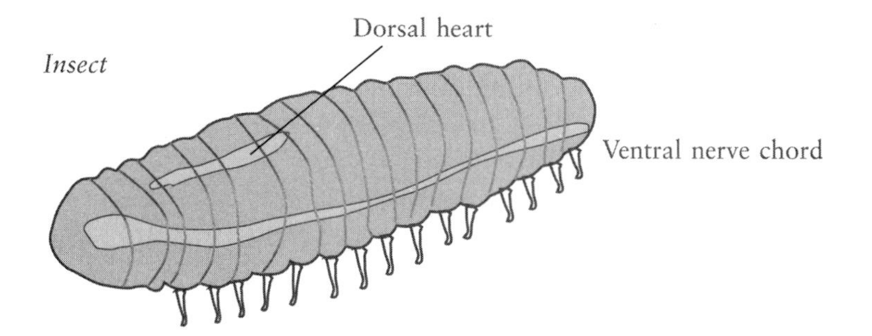

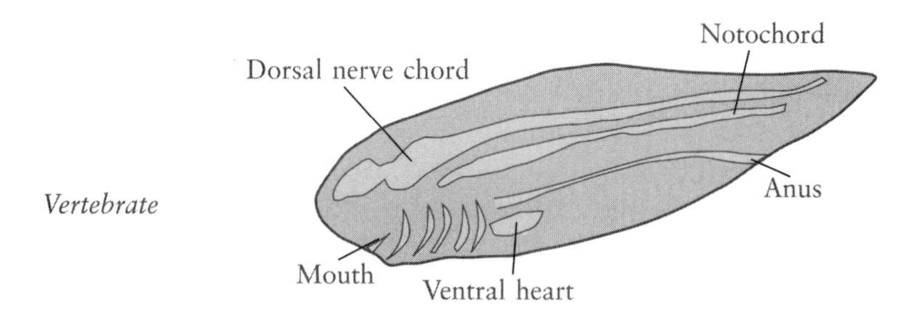

Body plans of complex bilateral animals that we know evolved in the Precambrian, 550 million years ago, and are seen in fossils.

'make much progress as they can neither see nor hear'.

In 1881 John Murray published Darwin's last book, *The Formation of Vegetable Mould Through the Action of Worms With Observations on their Habits* and it sold much better than anyone had expected, especially Murray, who reacted by boasting to one of his staff, 'Three thousand five hundred worms!' In the book's preface the author wrote 'The subject may appear an insignificant one' but to him it demonstrated an important principle about 'small agencies and their accumulated effects'.

Perhaps the book was popular because it conveyed the gentle pace of Victorian country life. This encompassed the changes of nature itself: the generation of new features and the regeneration of old ones, the selection of new characters and the exchange of others. Claude Bernard believed that when they are all together, living things became a system in equilibrium: each organism had an internal harmony, a self-organized order. Bernard was more interested in the work of the eighteenth-century scientist Joseph Priestley and the mathematician Pascal than anything to do with worms, but in 1879 Francis Galton wrote to his cousin to encourage him to consider the importance of Bernard's ideas when reflecting upon the question of an organism's state of equilibrium. Darwin didn't respond. The baton was not taken up until the start of the twentieth century by the Scottish polymath D'Arcy Wentworth Thompson.

D'Arcy Thompson had a background in classics, trained to be a doctor, loved mathematics and became professor of zoology at St Andrews at the age of twenty-four. All these experiences came together in his 1917 book *On Growth and Form*, which attempted to explain biology with physics, pushing Darwin's ideas of natural selection into the sidelines. In Thompson's view there was no need for inheritance when 'the world of living things, like the world of things inanimate, grows of itself, and pursues its ceaseless course of creative evolution'. Few biologists took him seriously. With physics he had disparaged heredity and historicism so that the Tree of Life was irrelevant.

Thompson described biological shapes in terms of coordinate geometry, arguing that their shapes were self-generated. Structure could evolve by direct imposition or by physical forces from the outside such as the determination of the shape of cells by their immediate neighbours. He liked to measure things, set their coordinates, and show how form could take shape from within the system itself. Structures like the single cells of protists and the much bigger shape of cauliflowers fascinated him for the opportunities they gave to study form mathematically.

Such ideas went down well in the intellectual circles of Scottish egalitarianism and his work on the beauty in the form of birds and fish were well appreciated by his old friends who had become scholars of classical Greek. But in the world of biology he became an isolated figure, admired for a beautifully written book with some clever observations of shapes, but criticized for a lack of attention to the way features are transmitted from one generation to the next. It cut him off from the scientists south of the border with England.

A hundred years later Thompson's ideas returned. At the Santa Fe Institute one of the few polymaths of the late twentieth century was searching for laws of complexity. Stuart Kauffman was the institute's director during the 1990s and, like D'Arcy Thompson, had started his career as a medic and then become a biophysicist. He is well known for his book *Origins of Order: Self-Organization and Selection in Evolution* (1993), which used the catchphrase 'Order for Free' to make clear the complex systems that make up biology. His ideas placed evolutionary biology back in the hands of the physicists but this time they are also listening to people in other disciplines including the geneticists. They also studied things like the florets of cauliflowers, finding their symmetry repeated the same intricate shapes at different levels of resolution, and was reminiscent of the shapes Thompson drew from his geometric logic. But Kauffman's analysis of many of these shapes showed a mathematical form very similar to one found by physicists studying self-organized systems.

Kauffman looked at many different kinds of self-controlling physical systems that act in both stable and stressed environments. The work these autonomous agents do, he found, through processes such as respiration and photosynthesis, drives the whole physiologically active body forwards. This means that in a new environment the organisms reach a particular equilibrium of activity. They do it by competition and selection, and if these many different things keep in balance the population glides forward confidently. Discretely, Kauffman and his colleagues were coming to the same conclusions that Lyell had with his ideas on uniformity, Spencer on competition and Darwin on natural selection. With complexity, they argued, each process acted to drive the system forward of its own accord, maintaining a balance on the edge of order and chaos, like a skater balancing their blades on ice with their own momentum. With any slight upset the self-regulating mechanisms are spurred to move forward and produce new origins and relationships, even go off into new directions.

From this vantage point, when the selection process is part of such a massive whole, the role of individual organisms in the big machines of society and ecology becomes unimportant. Like a single grain becoming dislodged to cause an avalanche on a sand pile, like an individual bee in a swarm, there is nothing special about any of these individuals. They just happen to be together in one place at one time; if it is not one it will be another. This is the perfect explanation for the entangled bank in the hedgerow up to Downe; how a growing sand pile cascades with unpredictable avalanches; or why a motor car repeatedly moves and stops in a traffic jam. Measurements of these waves of change show a standard and proven pattern. It shows up in all other self-organized systems and obeys rules that have been described as equations by physicists such as Per Bak. They are phenomena similar to the processes involved in Bernard's theory of homeostasis, another model describing how nature reaches a state of equilibrium on its own.

Within such a state of calm nothing is fixed, yet an air of tranquillity fills the scene. Well-organized logistics lie beyond the ability of the human gardener, in a place where there is no control or need to conform. There is no sense of prediction or form of design, beyond some simple rules of physics and chemistry. And it is hard to measure that which is not fixed. Within the world of nature, experiments and gardens have very limited ends.

15

The Unweeded Garden

Death at Down

At Down House, within only a few days of one another in April 1882, Charles Darwin and his daughter's dog Polly both died. Polly was laid to rest under an apple tree in the orchard, a spot not a mile away from the St Mary's churchyard where her master had expected to be buried. The graveyard held the remains of Emma and Charles's two young children and his brother Erasmus who had died the year before.

Members of the X Club had other ideas on Charles's resting place however and they persuaded the family as well as the Prime Minister and the Dean that he should be buried in Westminster Abbey. Joseph Parslow, who had worked as the butler since Emma and Charles married in 1839, spoke for the villagers at Downe: 'There was great disappointment in Downe that he was not buried there. He loved the place and we think he would rather have rested there had he been consulted.'

Instead, the coffin was carried by the Duke of Argyll, the Duke of Devonshire and the Earl of Derby, together with the American ambassador, the president of the Royal Society, Hooker, Wallace, Huxley and Lubbock. Darwin had not been interested in monuments; no permanent portrayal of life would do for him. Even his last book, the one about worms, considered the powers of digestion and how the life of the world is dependent on death. One of the things that he liked about worms was their collaborative generosity, whereby the earth was reused again and again, passing through their bodies, sustaining the fertility of the planet. It was, for Darwin, a world in which everything vanishes in the end, every individual has losses to mourn, and where altruism manifests itself in many, often devious, ways.

Yet it is tempting to seize upon mourning as a permanent thing, to refuse change, to cling to lost things, hoping to be something else, somewhere else. One who held this view was the God-fearing author of the *The Natural History of Selbourne* (1789), Gilbert White, who also found earthworms 'a small and despicable link in the chain of nature, yet, if lost, [they] would make a lamentable chasm'.

There was another chain that Darwin lamented from his own life and he had been wise enough to realize it. In May 1881 he wrote in the final pages of his autobiography:

> I have said that in one respect my mind has changed during the last twenty or thirty years. Up to the age of thirty, or beyond it, poetry of many kinds, such as the works of Milton, Gray, Byron, Wordsworth, Coleridge, and Shelley, gave me great pleasure, and even as a schoolboy I took intense delight in Shakespeare.

Then something changed:

> But now for many years I cannot endure to read a line of poetry: I have tried lately to read Shakespeare, and found it so intolerably dull that it nauseated me. I have also almost lost my taste for pictures or music . . . My mind seems to have become a kind of machine for grinding general laws out of large collections of facts, but why this should have caused the atrophy of that part of the brain alone, on which the higher tastes depend, I cannot conceive.

Like Darwin, Hamlet had a hunch about what had gone before,

and like Darwin he had become afraid in that isolation, knowing a truth when all around were unaware:

> How weary, stale, flat, and unprofitable
> Seem to me all the uses of this world!
> . . . 'tis an unweeded garden,
> That grows to seed; things rank and gross in nature
> Possess it merely.

There was more to nature than the sum of its constituent parts, something about its whole, its sense of order and forwardness.

Altruism at Down House

Seven of Emma and Charles's ten children went on to live into the twentieth century and continued the family legacy of bright individualism. Etty stood out as a particularly fastidious woman, according to one of her nephews:

> Fussy people Darwins are
> Who's the fussiest by far?
> Several aunts are far from calm,
> But Aunt Etty takes the palm.

Henrietta Darwin, or Etty for short, was born in September 1843, a year after Charles and Emma moved into Down House. In her old age she became a shrunken rather frail woman, a hypochondriac with no children of her own. Every day Etty was served breakfast in bed and was attended by maids and servants. Her isolation from the outside world soon developed into a series of eccentric habits, such as her own inventions of ways to keep off stray germs and how to avoid catching other peoples' illnesses. Her tea-strainer face-mask was filled with wadding soaked in eucalyptus oil and was held on by elastic bands. Sometimes she preferred to use the whole-body cover with its particularly ingenious fumigation system. Once, unhappy with the view of the garden from bed, she sent for the gardener to come up to her room. She told him to get into bed and work out which branch of the tree in front of the window needed cutting back so she could see the garden.

It was a life of high order and routine and even the illnesses were recorded – in the family bible. This was not used for much else and at one time was kept in a vault at the bank. Etty vilified religion and was especially anti-Catholic. 'I could swallow the Pope of Rome, but what I can NOT swallow is the celibacy of the clergy' she protested to her tired niece after an evening's discussion about religion.

It seemed that Etty had little of the selfless concern for the well-being of others that her father had suggested was an evolving feature of social animals. He proposed that our social instincts are part of the evolutionary process, giving us such behaviour as sympathy, kindness and the desire for social approbation. But they were very different to the structural characters associated with units of inheritance, so the question was whether these patterns of behaviour were inherited and, if so, in what form? Many species of social animals show characteristics of social cooperation, reciprocity and other forms of mutual help, not only in hunting and gathering but also in joining others to fight offensive groups and species. Human aggression can be especially strong, especially when men join national and religious armies.

Aggression also shows up in other ways, as in rebellion against the standards of a group, and Etty offered no exception, an extreme individual in a species whose society placed altruism as a major feature in its evolution. This altruism is usually strong in situations of low fear, such as in family environments, when it allows cooperation and sharing. But when trust breaks down, most individuals revert to conflict as a form of defence. Game theorists like Maynard Smith call this 'tit for tat': cooperation gets less and less until it gives way to retaliation, fighting beginning out of a fear that the other will attack first.

In humans these reactions are reinforced by psychological identification with recognized groups. We proudly call ourselves things like scientists, artists, Muslims, Jews or bankers, which often become the start of divisions between 'us' and 'them'. Poor Etty had no such web of trust and self-regard as she spent her life in isolation from mainstream society. But she wasn't entirely devoid of that other important evolutionary attribute of altruism: sexual feeling.

During the autumn in the copse at Down there were usually good crops of the woodland mushroom *Phallus impudicus.* Instead of blowing their spores through the air these stinkhorn mushrooms produce a sticky spore mass on their tip. The foul smell attracts flies which get the spores on their legs and inadvertently carry them to other locations. Shocked that 'the morals of the maids' might be influenced by the sight and smell of these hardcore objects, Etty performed a seasonal sport. Dressed in a red dressing gown, several shawls, scarves and rugs, and equipped with rubber hot-water bottles and boots, her hair in a pigtail and without her teeth, she approached the copse with a special hunting cloak and gloves. Stick in hand, she spiked the putrid mushrooms through their erect pale stipes. After each of these forays Etty secretly took her catch to the kitchen stove and burnt the phallic objects.

Of course, this was not how sexual reproduction worked. For these fungi the spores were transported long distances on the flies' feet, then left behind to bud more individuals in the right conditions. The simple mechanism of asexual reproduction, that is, without male fertilization of the female, had been the subject of experiments Etty's father carried out, some disguised as games for the children who shared his own genes. Their games with worker bees involved 'five or six children each close to a buzzing place' with Darwin telling 'the one farthest away to shout out "here is a bee" as soon as one was buzzing around'. Then the next in line of the bee's flight would shout and the pattern of the insect's flight was meant to emerge. Like so many of the experiments in the garden nothing much came out of it and the mysteries of the bee's flight and the altruistic castes of the insect remained.

Despite Etty's unusual habits her relations were a very loving family. At home, once when their father was ill in bed, Horace took up to him the old *Beagle* telescope so Darwin could look at what was going on in the garden. Leonard was involved with work in the greenhouse, while Frances took over many of the plant experiments in the garden. Emma was warm and competent, an inspiring wife and mother and, as was the way in Victorian households with educated parents and governesses, a lot of time was spent reading aloud and playing the piano. Between their summer walks and picnics the middle classes also had time to play croquet and tennis.

Group behaviour in evolution

After the Darwin family left for Cambridge with Emma's death in 1896, Down House became a school, then the school moved on and it became a museum run by the British Association. In the closing months of the Second World War, the country around Down was in the flight path of the buzz bombs that droned from across the English Channel to fall on the London suburbs. After the war, visitors returned to look at the house and gardens. One of them, on a day trip with his mother from their home just a few miles away, was a ten-year-old boy called Bill Hamilton. He was interested in insects and had just been given a copy of *Butterflies* by the fastidious Oxford bachelor E. B. Ford, author of the first of the New Naturalist series.

Hamilton went on to study genetics at Cambridge in the 1950s and then at London. He ended up in Ford's zoology department at Oxford where he became one of the century's greatest evolutionary biologists. As well as knowing a great deal about social insects, those that swarm and work together, Hamilton liked to use mathematics to take the patterns he found in their behaviour a stage or two further. His background in genetics allowed him to follow the intellectual lineage from Francis Galton, through the multivariate statistics of Carl Pearson, the mathematics of J. B. S. Haldane to the game theory of John Maynard Smith. In their different ways and at different times these four men had been trying to find a mathematical equation to describe natural selection. These were stern and hard men to follow, but Hamilton's gentle manner was not perturbed by their bullying. His courage paid off for it was he who won through to give evolutionary biology the different kind of understanding they had been seeking. Hamilton was a remote kind of man and rarely mixed with people, and his unawareness of their conventions got him into trouble with Maynard Smith. Yet the intellects of these strong scientists were on the edge of creative validity and it was inevitable that from time to time they should cross conventional lines.

What Hamilton found was that where an ecosystem helps relatives to be together, then altruism will evolve. It meant that a behavioural character like kinship could be ranked alongside anatomical and biochemical characters as a factor in survival and evolution. What's more, Hamilton expressed this simple balance of relatedness, benefit and cost by the equation $rb - c > 0$. After initial disbelief following its publication in 1964, the equation has been tested many times on data from social insects, primates and other organisms and has been found to work.

One familiar example to demonstrate Hamilton's work is from the many children's stories about rabbits running around and playing with one another. When the big bad wolf comes along, the first rabbit to notice jumps up and warns all the others by showing the white patch under its tail. Instantly all the rabbits show their white bobtails and run off. These are no special rabbits, but they are related, all members of the same family – they have evolved a simple mechanism of self-protection with clear benefits from being in a good place with a large group. Of course, if a rabbit fails to notice, fails to join in with the group, the big bad wolf will get him.

Another feature of group behaviour was what Galton, in the late nineteenth century, called the 'middle-most' value. This came from all parts of the group and explained how an average trend was found across nature. The range of human intelligence was one of the first things to be tested in this way and Galton devised original ways of measuring it which showed the variation in human intelligence more usefully as a bell-curve. These topics were then being raised for the first time as problems for biology rather than for theology. Another issue was whether sterile altruistic castes of worker bees were created by the divine hand of the Creator or had evolved because they were profitable for others in the hive.

Similar kinds of arguments covered many other topics throughout the twentieth century and some stay with us, mostly on the cusp between science and art. The hardline view was led by a school of scientists looking into a cell's chemistry and expecting to measure it. The work began with Galton and then went from Haldane to Maynard Smith and Hamilton.

A parallel group with a soft outlook considered what went on outside the cells and organisms, and the organism's relationship with the environment. Here were people such as Arthur Tansley, the founder of ecology in Britain, but they didn't make much of a mark in wartorn Europe. All their interactions were dominated by the stresses of society around them, the sudden catastrophes of war, not the steady habits of altruism.

Hardliners at Oxford and the key to life

In the 1970s and 1980s ever inquisitive scientists were measuring more phenomena, from biology to economics, by looking for the smallest units that sustain a system, how they work and change. Mrs Thatcher was focusing on the smallest inner workings of the economy, giving power to the individual and arguing there is no such thing as society. Her monetarist policy fitted with the mood of those times and came from the focused analytical way of thinking that she had acquired when studying chemistry at Oxford. In 1976 Richard Dawkins took his colleague Bill Hamilton's mathematical analysis of altruism into the public domain with his book *The Selfish Gene,* whose thesis was that genes are *the* units of selection, and which quickly became a bestseller.

It was the mission of many of the hardline scientists to focus on what they thought were the smallest and most fundamental parts of the evolutionary mechanism. They expected to smooth a way out of the arguments going on outside the cell by giving clear rules from the building blocks inside. They hoped this approach would enable more gateways to open up, with more people speaking the same language of this fundamental evolutionary biology. Instead, as might have been expected in the holistic world of nature, full of very different bits and pieces that keep changing, their hardline policy has revealed more contradictions and differences than answers. The same accumulation of new questions goes on: try as they might, scientists cannot take away the ambiguity. It is why so few agree with one another.

One ambiguity in evolutionary biology comes from the hard-line search for an ever-reducible cause, what some call the 'gene', and the equally evasive level at which selection actually happens. Are these the same thing, in the same cell? Is there this one place where selection, adaptation or some part of evolutionary change actually happens, once for each advance? Or does selection happen between cells or organs, individuals or families, or even species?

These modern debates are re-enactments of old Victorian arguments, like those between T. H. Huxley and 'Soapy Sam' Wilberforce, in which the extremes of different kinds of scientist were set against the less-measured world of philosophers and theologians. Then they were polarized groups who had no trust in one another and expected to fight instead. Darwin's '*Man* book', however, drawing on his South American experiences and observations at Down House, gave several subtle suggestions that natural selection can happen not only at the level of an individual, but also as a family group and at the species level. Like rabbits with their tails, groups do what individuals and genes can't.

T. H. Huxley was one who was sceptical of this inward looking approach to the new way of thinking about life, no doubt influenced by other liberals. He had also been listening to Galton and Nietzsche. He later became influential himself on future mathematical scientists such as Haldane, but tended to urge a more cautious approach. In 1890 he said,

> mathematics may be compared to a mill of exquisite workmanship, which grinds you stuff of any degree of fineness; but nevertheless, what you get out depends upon what you put in it; and as the grandest mill in the world will not extract wheatflour from peascods, so pages of formulae will not get a definite result out of loose data. Darwin would have agreed.

Followers of those who proposed a scientific explanation of life were beginning to believe in science as a new religion, that it would break through natural processes and allow humans to take control. Nevertheless, they retained the old notion that our species is the final target of evolution. Stimulated by his Modern Synthesis, those around Haldane such as Maynard Smith encouraged the waves of thinking that led to Hamilton's equation and its mathematical proof. They were a new generation trying to find patterns from the evolutionary goings-on in nature. They argued together and inspired one another in their own interests, like a family fighting their fears from outside and from the unknown.

Many of the developments in the study of evolutionary biology that we have seen during the twentieth century made it appear that science was going to resolve the major mechanisms of evolutionary processes. Mendelian genetics and later molecular biology approached these problems on different scales and led to understanding right down to the smallest level. The work of hox genes is the latest breakthrough and who knows where the next level down will take us? There is no doubt that the hard-line scientists have been very successful in outlining these processes inside the cell.

Yet it is hard to share their optimism when you follow this story of Darwin's garden and its secrets, or at least those few that have unfolded over the last 150 years. That is because, still, the pathways of evolution are adapting to changes in environment and our species' social circumstances. As we think of the consequences for science, as we devise the next experiment, nature's opportunism is ready, waiting to react. Quickly it has passed us by, leaving us with huge databases too big to analyse, confusion about whether it is the genes or individuals or species that evolve, ignorance of how the big groups originated, whether they formed gradually or catastrophically. Each question is being answered with more questions. Nature is always ahead of science.

Return to nature

I walk from the main entrance of the Department of Biological Sciences at University College, a 1960s block on the site of the Darwins' house in Gower Street. Just round the corner is Euston

Station and then the Eurostar Terminal to Cuvier's Paris. I drive through the city and out of London. The Old Kent Road still leads from Borough Market south-east to the long slope up the river terraces and the chalk escarpment to Downe. There is still the narrow lane and, as I stop to let cars pass, there, in the hedgerow, I see the hedge parsley.

I take one of the many footpaths near Down House and find myself crossing the Sandwalk with views of the countryside still much the same as 150 years ago. The fields and woodlands still make the same shapes and colours on the folding hillsides and from this vantage point the wildlife gives an atmosphere of mystery and beauty. Darwin's house behind me is not lived in any more but is in the care of English Heritage and is furnished as it was in Darwin's day, with billiards and Frank's bassoon, Emma's grand piano and Charles's study. The tennis racquets have been put back under the stairs. The gardener gives tours to show the planting that goes on in the spring and leads fungal forays in the autumn. Still, the house is quiet as it has always been and the garden has its worms.

Back at my own home I search the Internet for details of the house and Darwin's papers. I find the text of a lecture that Thomas Huxley gave eleven years after Darwin's death that raised humans as a species above all others. Surprised at the imprecision and scope he left for interpretation, I find that Huxley's 1893 Romanes Lecture centred on the story of 'Jack and the Beanstalk' – the struggles of human instinct and desire on the one hand, the overpowering control of the undergrowth of nature on the other.

From the entangled bank on the road to Downe, to 'Jack and the Beanstalk' in Huxley's lecture, it is only a short way to the gardener struggling with the hostile forces of the natural world. Jack climbed up in search of his fantasies, vividly brought to life at the top by the giant and the beautiful princess. But gardens, too, are artificial places where humans wander into their own space, places of fashion and social change, as Huxley was clearly well aware. They are not supposed to be the same as the environments of nature where conflict and ambiguity reign over the weeds of unharmonious species.

In nature the changes can happen unpredictably and quickly, and take place on many different scales and disciplines and so they are enigmatic. At last, however the garden always reveals its open secret: nature will be taking back our garden. In Huxley's fairy story, the giant and the beautiful princess were mere fantasies; industry, empire and science were the accepted parts of their lives that they took for granted. And now we know that nature is devious enough to set its own level, like water in a stream running around into nooks and crannies, eventually through grains and banks of sand and shingle, out into the sea. It is too cunning for humans' scientific progress to monitor, let alone catch. Away from the competition of organic life, the environment with its inorganic landscape is stronger than humans. The garden will always return to the wilderness, the gardener's attempts to control its forces are temporary and superficial.

In their 1927 book, *Animal Biology*, J. B. S. Haldane and Julian Huxley, Thomas's grandson, expressed a strong belief, still shared by most people, that 'man has . . . the possibility of consciously controlling evolution according to his wishes'. To those of us who realize, now, that the environment is entering a new phase of another catastrophic event, these sound like hollow words.

Against this grand rhetoric Darwin's garden offers a response: at some point science's power will decline and nature will claim Down back.

Influences and Sources

The facts and feelings that are set out in this book have come together and been developed in my experiences of life since the one hundredth anniversary of the publication of *The Origin of Species*. Numerous teachers, colleagues and friends have been involved, especially my small family, and people at Alderman Newton's School Leicester, University College London, Imperial College London, the University of East London and the Natural History Museum. I have benefited from strong relationships with scores of other scientists through the International Organization of Palaeobotany and the International Union of Biological Sciences.

I would like to offer my special thanks to Leo Hollis of Constable & Robinson who made this project possible from the start; he helped devise the structure of the book and edited every page. The title which brought it all together was the idea of Peter Tallack of Curtis Brown. Other support and encouragement came from Jeremy Barlow, Ras Darwin, Robert Darwin, Raphie Kaplinsky, Colin Merton, Trevor Robinson, Osman Streater, Derek Winterbottom and Martin Wooster.

Most of Charles Darwin's writing is now freely available, and searchable, on the Internet.

Specialist help for Chapters 11 and 13 was kindly given by Richard Bateman, Bill Chaloner, Margaret Collinson and David Dilcher. Many Internet websites were consulted and some were used cautiously. The librarians at the Linnean Society, the

Geological Society and the Royal Society were always willing to search for difficult references.

General Sources and Influences

Bowlby, J., *Charles Darwin*, Hutchinson, London (1990).
Browne, J., *Charles Darwin Volume I: Voyaging*, Pimlico, London (1995).
Browne, J., *Charles Darwin Volume II: The Power of Place*, Pimlico, London (2002).
Corning, P. A., *Holistic Darwinism: Synergy, Cybernetics, and the Bioeconomics of Evolution*, University of Chicago Press, Chicago (2005).
Darwin, C., *The Origin of Species*, Penguin, London (1980 [1859]).
Darwin, F., *The Life and Letters of Charles Darwin, Including an Autobiographical Chapter*, 3 volumes, John Murray, London, (1887).
Darwin, F. and Seward, A. C., *More Letters of Charles Darwin: A Record of his Work in a Series of Hitherto Unpublished Letters*, John Murray, London (1903).
Desmond, A. and Moore, J., *Darwin: The Life of a Tormented Evolutionist*, Michael Joseph, London (1991).
Gould, S. J., *The Structure of Evolutionary Theory*, Harvard University Press, Cambridge, Mass. (2002).
Kohn, M., *A Reason for Everything: Natural Selection and the English Imagination*, Faber and Faber, London (2004).
Raverat, G., *Period Piece*, Faber and Faber, London (1952).
Rose, H. and Rose, S. (eds.), *Alas, Poor Darwin*, Cape, London (2000).

1. Down House

Ashton, R., *142 Strand*, Chatto and Windus, London (2006).
Atkins, H. A., *Down: The Home of the Darwins*, Royal College of Surgeons, London (1974).
King-Hele, D., *Erasmus Darwin: A Life of Unequalled Achievement*, Giles de la Mare, London (1999).

Le Guyader, H., *Geoffroy Saint-Hilaire: A Visionary Naturalist*, Chicago University Press, Chicago (2004).
Morris, S., Wilson, L. and Kohn, D., *Charles Darwin at Down House*, English Heritage, London (1998).
Quammen, D., *The Kiwi's Egg: Charles Darwin and Natural Selection*, Weidenfeld & Nicolson, London (2006).
Ruse, M., *Darwin and Design: Does Evolution Have a Purpose?*, Harvard University Press, Cambridge, Mass. (2003).
Uglow, J., *The Lunar Men*, Faber and Faber, London (2003).
White, G., *Natural History and Antiquities of Selbourne*, Ray Society, London, 1789.

2. A New Garden at Down House

Blunt, W., *Linnaeus: The Complete Naturalist*, Frances Lincoln, London (2004).
Gopnik, A., 'Rewriting nature: Charles Darwin, natural novelist', *The New Yorker*, 23 October 2006, pp. 50–59.
Healey, E., *Emma Darwin: The Inspirational Wife of a Genius*, Headline, London (2000).
Jarvis, C., *Order Out of Chaos: Linnean Plant Names and their Types*, Linnean Society of London (2007).
Keorner, L., *Linnaeus: Nature and Nation*, Harvard University Press, Cambridge, Mass. (1999).
MacGregor, A. (ed.), *Sir Hans Sloane*, British Museum, London (1994).
Mayr, E., *What Evolution Is*, Weidenfeld & Nicolson, London (2002).
Stafleu, F. A., *Linnaeus and the Linneans*, International Association for Plant Taxonomy, Vienna (1971).
Turner, J. S., *The Tinkerer's Accomplice: How Design Emerges from Life Itself*, Harvard University Press, Cambridge, Mass. (2007).

3. A Slow Start at Down

Keynes, R., *Annie's Box: Charles Darwin, his Daughter and Human Evolution*, Fourth Estate (2001).

Lyell, C., *The Geological Evidences of the Antiquity of Man, with an Outline of Glacial and Post Tertiary Geology, and Remarks on the Origin of Species, with Special Reference to Man's First Appearance on the Earth*, John Murray, London (1873).
Rudwick, M. J. S., *The Meaning of Fossils: Episodes in the History of Palaeontology*, Macdonald, London (1972).
Rudwick, M. J. S., *Bursting the Limits of Time*, Chicago University Press, Chicago (2005).

4. The Tree of Life

Cowan, R., 'Barcoding plants', *Kew Scientist*, 2005, Vol. 27, No. 1, p. 1.
Nee, S., 'The great chain of being', *Nature*, 26 May 2005, Vol. 435, p. 429.
Tudge, C., *The Secret Life of Trees*, Allen Lane, London (2005).

5. Entanglements

Davies, K. G., 'Creative tension: what links Aristotle, William Blake, Darwin and GM crops?', *Nature*, 14 September 2000, Vol. 407, p. 135.
Mabey, R., *Nature Cure*, Chatto & Windus, London (2005).
Nee, S. and Colegrave, N., 'Paradox of the clumps', *Nature*, 25 May 2006, Vol. 441, pp. 417–8.
O'Hear, A., *Beyond Evolution: Human Nature and the Limits of Evolutionary Explanation*, Clarendon Press, Oxford (1997).

7. Actions Out of Quietness

Brookes, M., *Extreme Measures and the Dark Visions and Bright Ideas of Francis Galton*, Bloomsbury, London (2004).
Thwaite, A., *Glimpses of the Wonderful: The Life of Philip Henry Gosse*, Faber and Faber, London (2002).

8. Exploring the Gradual

Barton, R., '"Huxley, Lubbock, and half a dozen others": professionals and gentlemen in the formation of the X Club, 1851–1864', *Isis*, September 1998, Vol. 89, pp. 410–44.
Gordon, D. M., 'Control without hierarchy', *Nature*, 8 March 2007, Vol. 446, p. 143.
Hendry, A. P., 'The power of natural selection', *Nature*, 17 February 2005, Vol. 433, pp. 694–5.
Hendry, A. P., 'The Elvis paradox: is stasis dead?', *Nature*, 8 March 2007, Vol. 446, pp. 147–50.
Lenski, R. E., Ofria, C., Pennock, R. T. and Adami, C., 'The evolutionary origin of complex features', *Nature*, 8 May 2003, Vol. 423, pp. 139–44.
Lovelock, J., *The Revenge of Gaia: Why the Earth is Fighting Back and How We can still Save Humanity*, Allen Lane, London (2006).
McCann, K., 'Protecting biostructure', *Nature*, 1 March 2007, Vol. 446, p. 29.
Moore, P. D., 'Where slugs may safely graze', *Nature*, 7 July 2005, Vol. 436, pp. 35–6.
Poelwijk, F. J., Kiviet, D. J., Weinreich, D. M. and Tans, S. J., 'Empirical fitness landscapes reveal accessible evolutionary paths', *Nature*, 25 January 2007, Vol. 445, pp. 383–7.
Raby, P., *Alfred Wallace*, Chatto & Windus, London (2001).
Stauffer, R. C. (ed.), *Charles Darwin's Natural Selection: Being the Second Part of his Big Species Book Written from 1856 to 1858*, Cambridge University Press, Cambridge (1975).
Strogatz, S. H., 'Romanesque networks', *Nature*, 7 January 2005, Vol. 433, pp. 365–6.
Theise, N. D., 'Now you see it, now you don't', *Nature*, 29 May 2005, Vol. 435, p. 1165.
Whitfield, J., 'Order out of chaos', *Nature*, 18 August 2005, Vol. 436, pp. 905–7.

9. The Holly and the Ivy

Boulter, M. C., Gee, D. and Fisher, H. C., 'Angiosperm radiations at the Cenomanian/Turonian and Cretaceous/Tertiary boundaries', *Cretaceous Research*, 1998, Vol. 19, pp. 107–12.

Emerson, B. C. and Kolm, N., 'Species diversity can drive speciation', *Nature*, 3 November 2005, Vol. 434, pp. 1015–7.

Gittenburger, E., Groenenberg, D. S. J., Kokshoorn, B. and Preece, R. C., 'Molecular trails from hitch-hiking snails', *Nature*, 26 January 2006, Vol. 439, p. 409.

Hooper, R., 'The scenic route out of Africa', *New Scientist*, 21 May 2005, p. 14.

Knight, D., 'Kinds of minds: whether scientists focus on pieces and particulars, or make broad connections', *Nature*, 10 May 2007, Vol. 447, p. 149.

Manen, J.-F., Boulter, M. C. and Naciri-Graven, Y., 'The complex history of *Ilex L.* (*Aquifoliaceae*): evidence from the comparison of plastid and nuclear DNA sequences and from fossil data', *Plant Systematics and Evolution*, 2002, Vol. 235, pp. 79–98.

Savolainen, V. *et al.*, 'Sympatric speciation in palms on an oceanic island', *Nature*, 11 May 2006, Vol. 441, pp. 210–13.

Whittaker, R. J., *Island Biogeography*, Oxford University Press, Oxford (1998).

10. The Rise and Fall of Mendel's Genetics

Arthur, W., *Biased Embryos and Evolution*, Cambridge University Press, Cambridge (2004).

Brakefield, P. M. and French, V., 'How and why to spot fly wings', *Nature*, 3 February 2005, Vol. 433, pp. 466–7.

Casti, J. L., *The Cambridge Quintet*, Abacus, London (1998).

Gelvin, S. B., 'Gene exchange by design', *Nature*, 10 February 2005, Vol. 433, pp. 583–4.

Henig, R. M., *The Monk in the Garden*, Mariner Books, New York (2000).

Huxley, J. (ed.), *The New Systematics*, Oxford University Press, Oxford (1940).

Moses, K., 'Fly eyes get the whole picture', *Nature*, 12 October 2006, Vol. 443, pp. 638–9.
Pearson, H., 'What is a gene?', *Nature*, 25 May 2006, Vol. 441, pp. 399–401.
Rosenfield, I. and Ziff, E., 'Evolving evolution', *The New York Review of Books*, 11 May 2006, pp. 12–15.
Stevens, C. F., 'Crick and the claustrum', *Nature*, 23 June 2005, Vol. 435, pp. 1040–41.

11. Orchids become Hopeful Monsters

Andrews, H. N., *The Fossil Hunters: In Search of Ancient Plants*, Cornell University Press, Ithaca (1980).
Arthur, W., 'The search for novelty', *Nature*, 17 May 2007, Vol. 447, pp. 261–2.
Bateman, R. M., Hilton, J. and Rudall, P. J., 'Morphological and molecular phylogenetic context of the angiosperms', *Journal of Experimental Botany*, 2006, Vol. 57, pp. 3471–503
Bateman, R. M. and Rudall, P. J., 'The good, the bad and the ugly: using naturally occurring terrata to distinguish the possible from the impossible in orchid floral evolution', *Aliso*, 2006, Vol. 22, pp. 481–96.
Biemont, C. and Vieira, C., 'Junk DNA as an evolutionary force', *Nature*, 5 October 2006, Vol. 443, pp. 521–4.
Bonetta, D. and McCourt, P., 'A receptor for gibberellin', *Nature*, 29 September 2005, Vol. 437, pp. 627–8.
Carroll, S. B., *Endless Forms Most Beautiful: The New Science of Evo-devo*, W. W. Norton & Co. Inc, New York (2005).
Cronk, Q. C. B., Bateman, R. M. and Hawkins, J. A. (eds.), *Developmental Genetics and Plant Evolution*, Taylor & Francis, Abingdon (2002).
Ke-Wei Liu *et al.*, 'Self-fertilisation strategy in an orchid', *Nature*, 22 June 2006, Vol. 441, p. 945.
Laubichler, M. and Maienschein, J., *From Embryology to Evo-Devo: a history of developmental evolution*, MIT Press, Cambridge, Mass. (2007).

Ledford, H., 'The flower of seduction', *Nature*, 22 February 2007, Vol. 445, pp. 816–7.
Rajkumari, J. D. and Longjam, R. S., 'Orchid flower evolution', *Journal of Genetics*, 2005, Vol. 84, pp. 81–9.
Ramirez, S. R., Gravendeel, B., Singer, R. B., Marshall, C. R. and Pierce, N. E., 'Dating the origin of the Orchidaceae from a fossil orchid with its pollinator', *Nature*, 30 August 2007, Vol. 448, pp. 1042–5.

12. Modern Ideas about Vertebrate Evolution

Chiappe, L. M., *Glorified Dinosaurs: The Origin and Early Evolution of Birds*, John Wiley & Sons, Bognor Regis (2007).
Chiappe, L. M. and Bertelli, S., 'Skull morphology of giant terror birds', *Nature*, 26 October 2006, Vol. 443, p. 929.
Cifelli, R. L. and Gordon, C. L., 'Re-crowning mammals: molecules versus morphology', *Nature*, 21 June 2007, Vol. 447, pp. 918-9.
Cotton, J. A. and Page, D. M., 'Going nuclear: gene family evolution and vertebrate phylogeny reconciled', *Proceedings of the Royal Society Biological Sciences*, London, 2002, Vol. 269, pp. 1555–61.
De Duve, C., 'The onset of selection', *Nature*, 10 February 2005, Vol. 433, pp. 581–2.
Fondon, J. W. and Garner, H. R., 'Molecular origins of rapid and continuous morphological evolution: dog domestication and the evolution of complex life', *Proceedings of the National Academy of Sciences USA*, Vol. 101, pp. 18058–63.
Gee, H., 'Careful with that amphioxus', *Nature*, 23 February 2006, Vol. 439, p. 923.
Goldsmith, T. H., 'What birds see', *Scientific American*, July 2006, pp. 48–57.
Holmes, R., 'Recreating dinosaur genes', *New Scientist*, 25 April 2005, pp. 42–5.
Jablonka, E. and Lamb, M., *Evolution in Four Dimensions*, Bradford Books, Cambridge, Mass. (2005).
Janvier, P., 'Modern look for ancient lamprey', *Nature*, 26 October 2006, Vol. 443, pp. 921–4.

Kirschner, M. W. and Gerhardt, J. C., *The Plausibility of Life: Resolving Darwin's Dilemma*, Yale University Press, London and New Haven (2005).

Patel, N. H., 'How to build a longer beak', *Nature*, 3 August 2006, Vol. 442, pp. 515–6.

13. 'A Most Perplexing Phenomenon'

Boulter, M. C. and Kvacek, Z., 'The Palaeocene flora of the Isle of Mull', *Special Papers in Palaeontology*, Vol. 42, Palaeontological Association (1989).

Dilcher, D. L., 'Towards a new synthesis of major evolutionary trends in the angiosperm fossil record', *Proceedings of the National Academy of Sciences USA*, Vol. 97, pp. 7030–6.

Feild, T. S., Arens, N. C., Doyle, J. A., Dawson, T. E. and Donoghue, M. J., 'Dark and disturbed: a new image of early angiosperm ecology', *Paleobiology*, 2004, Vol. 30, pp. 82–107.

Friedman, W. E., '*Amborella trichopoda* from New Caledonia: last remnant of an ancient angiosperm lineage', *Nature*, 18 May 2006, Vol. 441, pp. 337–40.

Friis, E. M., Pedersen, K. R. and Crane, P. R., 'When Earth started blooming: insights from the fossil record', *Current Opinion in Plant Biology*, 2005, Vol. 8, pp. 5–12.

Frohlich, M. W., 'Recent developments regarding the evolutionary origin of flowers', *Advances in Botanical Research*, 2006, Vol. 44, pp. 63–127.

Moles, A. T., Ackerley, D. D., Webb, C. O., Tweddle, J. C., Dukia, J. B. and Westoby, M., 'A brief history of seed size', *Science*, 2005, Vol. 307, pp. 576–80.

Sun, G., Qiang, J., Dilcher, D. L., Zheng, S., Nixon, K. C. and Wang, X., '*Archaefructaceae*, a new basal angiosperm family', *Science*, 2002, Vol. 296, pp. 899–904.

14. Origins of Life

Allen, J. F. and Martin, W., 'Out of thin air', *Nature*, 8 February 2007, Vol. 445, pp. 610–12.

Bengtson, S., 'A ghost with a bite', *Nature*, 13 July 2006, Vol. 442, pp. 146–7.
Chourrout, D. *et al.*, 'Minimal ProtoHox cluster inferred from bilaterian and cnidarian Hox complements', *Nature*, 10 August 2006, Vol. 442, pp. 684–7.
Conway-Morris, S., *Life's Solution: Inevitable Humans in a Lonely Universe*, Cambridge University Press, Cambridge (2003).
Eichinger, L. *et al.*, 'The genome of the social amoeba *Dictyostelium discoideum*', *Nature*, 5 May 2005, Vol. 435, pp. 43–54.
Gray, M. W., 'The hydrogenosome's murky past', *Nature*, 3 March 2005, Vol. 434, pp. 29–30.
Kause, J. and Ruxton, G. D., *Living in Groups*, Oxford University Press, Oxford (2000).
Knoll, A. H., *Life on a Young Planet: The First Three Billion Years of Evolution on Earth*, Princeton University Press, Princeton, New Jersey (2003).
Nee, S., 'More than meets the eye', *Nature*, 24 June 2004, Vol. 429, pp. 804–5.
Pace, N. R., 'Time for a change', *Nature*, 18 May 2006, Vol. 441, p. 289.
Pilcher, H., 'Back to our roots', *Nature*, 23 June 2005, Vol. 435, pp. 1022–3.
Schmidt, K., 'Primeval pools', *New Scientist*, 2 July 2005, pp. 40–43.
Venter, C., 'Sea of genes', *New Scientist*, 14 May 2005, p. 21.

15. The Unweeded Garden

Armstrong, K., *The Great Transformation: The World in the Time of Buddha, Socrates, Confucius and Jeremiah*, Atlantic Books, London (2006).
Boulter, M. C., *Extinction: Evolution and the End of Man*, Fourth Estate, London (2003).
Bowler, P. J., *The Eclipse of Darwinism: Anti-Darwinian Evolution Theories in the Decades Around 1900*, Johns Hopkins University Press, Baltimore (1983).
Couzin, I. D., Krause, J., Franks, N. R. and Levin, S. A., 'Effective

leadership and decision-making in animal groups on the move', *Nature*, 3 February 2005, Vol. 433, pp. 513–16.
Damasio, A., 'Brain trust', *Nature*, 2 June 2005, Vol. 435, pp. 571–2.
Dawkins, R., *The Selfish Gene*, Oxford University Press, Oxford (1976).
Dawkins, R., *Unweaving the Rainbow*, Allen Lane, London (1998).
Dugatkin, L. A., *The Altruism Equation: Seven Scientists Search for the Origins of Goodness*, Princeton University Press, Princeton, New Jersey (2006).
Fodor, J., 'The selfish gene pool: Mother Nature, Easter bunnies and other common mistakes', *The Times Literary Supplement*, 29 July 2005, pp. 3–6.
Fukuyama, F., *The End of History and the Last Man*, Hamish Hamilton, London (1992).
Fuller, S., *Kuhn vs Popper*, Icon Books, Cambridge (2003).
Hamilton, W. D., *Narrow Roads of Gene Land*, W. H. Freeman, New York (1996).
Hull, D. L., *Science as a Process: An Evolutionary Account of the Social and Conceptual Development of Science*, Chicago University Press, Chicago (1988).
Midgley, M., *Science and Poetry*, Routledge, Abingdon (2001).
Midgley, M., *The Myths We Live By*, Routledge, Abingdon (2003).
Nugent, T., 'Darwin's disciple', *Amherst Magazine*, 2002, www.amherst.edu/magazine/issues/02fall/.
Ohtsuki, H., Hauert, C., Lieberman, E. and Nowak, M. A., 'A simple rule for the evolution of cooperation on graphs and social networks', *Nature*, 25 May 2006, Vol. 441, pp. 502–5.
Palla, G., Barabasi, A.-L., and Vicsek, T., 'Quantifying social group evolution', *Nature*, 5 April 2007, Vol. 446, pp. 664–7.
Phillips, A., *Darwin's Worms*, Faber and Faber, London (1999).
Queller, D. C., 'To work or not to work: coercion or kinship', *Nature*, 2 November 2006, Vol. 444, pp. 42–3.
Richerson, P. J. and Boyd, R., *Not By Genes Alone: How Culture Transformed Human Evolution*, Chicago University Press, Chicago (2005).

Scruton, R., 'Man and superman', *Prospect Magazine*, April 2003, www.prospect-magazine.co.uk.
Surowiecki, J., *The Wisdom of Crowds: Why the Many are Smarter than the Few*, Little, Brown, London (2004).

Index

CD Charles Darwin
DH Down House
Entries in **bold** indicate illustrations